Efficiency of Wireless Networks: Approximation Algorithms for the Physical Interference Model

Efficiency of Wireless Networks: Approximation Algorithms for the Physical Interference Model

Olga Goussevskaia

ETH Zurich
Switzerland
golga@tik.ee.ethz.ch

Yvonne-Anne Pignolet

IBM Research Zurich Laboratory
Switzerland
yvo@zurich.ibm.com

Roger Wattenhofer

ETH Zurich
Switzerland
wattenhofer@tik.ee.ethz.ch

the essence of knowledge

Boston – Delft

Foundations and Trends® in Networking

Published, sold and distributed by:
now Publishers Inc.
PO Box 1024
Hanover, MA 02339
USA
Tel. +1-781-985-4510
www.nowpublishers.com
sales@nowpublishers.com

Outside North America:
now Publishers Inc.
PO Box 179
2600 AD Delft
The Netherlands
Tel. +31-6-51115274

The preferred citation for this publication is O. Goussevskaia, Y.-A. Pignolet and R. Wattenhofer, Efficiency of Wireless Networks: Approximation Algorithms for the Physical Interference Model, Foundation and Trends® in Networking, vol 4, no 3, pp 313–420, 2009

ISBN: 978-1-60198-404-3

Foundations and Trends® in Networking

Volume 4 Issue 3, 2009

Editorial Board

Editorial Scope

Foundations and Trends® in Networking will publish survey and tutorial articles in the following topics:

- Ad Hoc Wireless Networks
- Sensor Networks
- Optical Networks
- Local Area Networks
- Satellite and Hybrid Networks
- Cellular Networks
- Internet and Web Services
- Protocols and Cross-Layer Design
- Network Coding
- Energy-Efficiency Incentives/Pricing/Utility-based
- Games (co-operative or not)
- Security
- Scalability
- Topology
- Control/Graph-theoretic models
- Dynamics and Asymptotic Behavior of Networks

Information for Librarians

Foundations and Trends® in Networking, 2009, Volume 4, 4 issues. ISSN paper version 1554-057X. ISSN online version 1554-0588. Also available as a combined paper and online subscription.

Foundations and Trends® in
Networking
Vol. 4, No. 3 (2009) 313–420
© 2010 O. Goussevskaia, Y.-A. Pignolet and
R. Wattenhofer
DOI: 10.1561/1300000019

Efficiency of Wireless Networks: Approximation Algorithms for the Physical Interference Model

Olga Goussevskaia[1], Yvonne-Anne Pignolet[2] and Roger Wattenhofer[3]

[1] ETH Zurich, Switzerland, golga@tik.ee.ethz.ch
[2] IBM Research Zurich Laboratory, Switzerland, yvo@zurich.ibm.com
[3] ETH Zurich, Switzerland, wattenhofer@tik.ee.ethz.ch

Abstract

In this monograph we survey results from a newly emerging line of research that targets algorithm analysis in the physical interference model. In the main part of our monograph we focus on wireless scheduling: given a set of communication requests, arbitrarily distributed in space, how can these requests be scheduled efficiently? We study the difficulty of this problem and we examine algorithms for wireless scheduling with provable performance guarantees. Moreover, we present a few results for related problems and give additional context.

Contents

1

Introduction

Despite the omnipresence of wireless networks, their fundamental communication limits are not fully understood: designing and operating a wireless network is often a matter of trial-and-error, regardless of whether it is a Wireless LAN in an office building, a GSM phone network, or a sensor network on a volcano.

We are interested in the fundamental communication limits of wireless networks. Given an *arbitrary* wireless network, and an *arbitrary* traffic pattern, we want to utilize the full bandwidth of our network. One of the most challenging characteristics of wireless networks is the fact that mutual interference impairs the quality of signals received and might even prevent the correct reception of messages. Efficient algorithms that coordinate the transmissions are therefore essential for the operation of wireless networks. To this end, we want to understand the maximum possible spatial reuse, i.e., which devices can transmit concurrently, without interfering. Given a set of communication requests, what is the minimum time needed to schedule all these requests successfully? How should media access be organized in a given network? In an existing wireless network, is it sensible to add relays, and where are they to be placed?

Evidently, if one hopes for analytic answers to questions like these, one must first decide for a reasonable wireless transmission model. In the past, a large fraction of analytic research on wireless networks has focused on models where the network is represented by a graph. The wireless devices are mapped to nodes and any two nodes within communication (or interference) range are connected by an (annotated) edge. Such graph-based models are particularly popular among higher-layer protocol designers, hence they are also known as *protocol models*. Unfortunately, protocol models are often too simplistic: consider, for instance, a case of three wireless communication pairs, every two of which can be transmitting concurrently without a conflict. In a protocol model, one will conclude that all three transmissions may transmit concurrently as well, while in reality this might not be the case since wireless signals accumulate. Instead, it may be that any two transmissions together generate too much interference, hindering the third receiver from correctly receiving the signal of its sender. This many-to-many relationship makes understanding wireless transmissions difficult; a model where interference accumulates seems paramount to truly comprehending wireless communication. Similarly, protocol models oversimplify wireless attenuation. In protocol models the signal is usually "binary", as if there was an invisible wall at which the signal ends abruptly. Not surprisingly, in reality the signal strength decreases gracefully with distance. Because of these shortcomings, results for protocol models are often not applicable in reality.

In contrast to the algorithmic ("computer science") community which focuses on protocol models, researchers in information, communication, and network theory ("electrical engineering") are working with wireless models where interference accumulates and attenuation is taken into account. A standard model is the physical model; we will formally introduce it in Section 2. In this model, the energy of a signal fades with the distance to the power of the path-loss parameter α. If the signal strength received by a device divided by the strength of interference caused by concurrent transmitters (plus the noise) is above some threshold β (signal-to-interference-plus-noise ratio (SINR)), the receiver can decode the message, otherwise it cannot.

Unfortunately, most work using the physical model does not provide algorithms with provable performance guarantees. Usually heuristics are proposed instead, evaluated by simulation. Analytical work is done for special cases only, e.g., networks with a grid structure, or random traffic. However, these special cases do not give much insight into the complexity of the problem; also, it seems difficult to derive new protocols from analytical work on special cases. If one is interested in the capacity of an *arbitrary* wireless network, and how this capacity can be achieved, an algorithmic approach seems unavoidable.

In this monograph we present recent results that combine the best of both worlds: we present algorithms and bounds for arbitrary wireless networks (*not* random node distributions), using the physical model (*not* the protocol model). We believe that bridging the gap between protocol designers and communication theorists is a fundamental challenge of the coming years, a hot topic for the wireless network community with implications for both theory and practice. To the best of our knowledge, research in this emerging area is only a few years old [66]. Nevertheless, the body of work is growing rapidly. Hence we cannot provide a complete survey; instead we focus on wireless scheduling using a simple physical model. More precisely, given a set of communication requests, arbitrarily distributed in space, how can these requests be scheduled efficiently? This question may be formulated in several ways, using different parameters. One might want to know the maximum number of requests that can be scheduled simultaneously. Alternatively, one might ask what is the minimum time needed to schedule all requests. Essentially, the main objective is to achieve efficient spatial reuse, considering wireless interference among nodes transmitting concurrently. Such results promise to lead to answers to questions such as "What is the throughput capacity of a specific wireless network?", and "*How* can this capacity be realized?"

This monograph is organized as follows: In Section 2 we formally define the models and problems of interest; in addition we present a robustness result that shows that small perturbations in the model do not fundamentally change the results. The main content is in Sections 3 and 4. In Section 3 we study wireless scheduling *without* power control, and in Section 4 *with* power control. As we will see, most of the questions are NP-hard, so we settle for so-called approximation algorithms, algorithms that guarantee that a solution is at most a bounded factor worse than optimum. We focus on simple (and to some degree teachable) results, and usually merely mention more elaborate techniques. In Section 5 we will survey a few results beyond scheduling. Finally, in Section 6 we provide additional context about related areas.

At the time of writing, results are emerging that reconsider problems and results for protocol models successfully in the physical models. Indeed, this direction of research is increasingly popular, as first surveys and overview articles [62] are published. Analogously, we hope that some of the ground-breaking research on special-case topologies in the physical model may be generalized and studied in an algorithmic way.

2

Models and Definitions

When studying scheduling in wireless networks, the choice of an appropriate interference model is crucial. In the past, researchers have studied a wide range of interference models, ranging from complex physical models to simple graph-based protocol models.

In this chapter we introduce the main wireless interference model used to derive the results presented in this survey, namely, the physical interference model. Subsequently, we define in Section 2.2 three main problems we address in this monograph: the One-Slot Scheduling Problem, the Weighted One-Slot Scheduling Problem, and the Multi-slot Scheduling Problem.

The chapter is concluded in Section 2.4, where an analysis of robustness properties of the physical interference model with respect to parameters, such as signal strength and spatial dispersion, is presented. This section also contains notation and definitions, important to understand the results presented in Sections 3 and 4.

2.1 Physical Interference Model

In this survey we study the problem of scheduling wireless communication requests (or simply links) in the *physical interference model* [38].

5

We assume that the *input* to the problem is a set $L = \{\ell_1, \ldots, \ell_n\}$ of n wireless links, where each link ℓ_i represents a communication request from a sender s_i to a receiver r_i:

$$\ell_i = (s_i, r_i).$$

The communication devices are viewed as nodes positioned in a Euclidean space. The distance between two nodes s_i, r_j is denoted by $d_{ij} = d(s_i, r_j)$, so the length of a link ℓ_i is referred to by $d_{ii} = d(s_i, r_i)$. We assume that there are no primary conflicts in the transmission setup, i.e., each node is either a sender or a receiver and each receiver is associated with only one sender. Scenarios with this type of conflicts can be reformulated by introducing additional nodes at the same position, i.e., if a receiver is associated with two senders, two links with co-located receivers can be used, resulting in one sender–receiver pair for each link. Therefore, we neglect these scenarios for simplicity's sake.

Moreover, we assume that each link has a unit-traffic demand, and model the case of *non-unit traffic demand* by replicating each link k times, where k is the demand on the link.

In the *physical interference* model, the received signal power decays proportionally to the inverse of the distance between the sender and the receiver to the power of a so-called *path-loss exponent* α, which is a constant, whose exact value depends on external conditions of the medium (humidity, obstacles, etc.), as well as the exact sender–receiver distance. The faster the signal strength falls, the smaller the amount of interference caused. In Ref. [74], measurements of indoor and outdoor path-loss exponents at various frequencies are reported, ranging from 1.6 to 6. Most work relies on the assumption that $\alpha > 2$, exploiting the fact that in this case the interference of far away nodes can be bounded easily. The propagation attenuation or *link gain* between sender s_i and receiver r_i is therefore $d_{ij}^{-\alpha}$. If $P(s_i)$ is the transmitting power level of a sender s_i, the received power at the receiver r_i is:

$$P_{ii} = P_{r_i}(s_i) = \frac{P(s_i)}{d_{ii}^{\alpha}}.$$

The power received from s_i by the receiver r_j of a concurrently scheduled link ℓ_j is referred to as *interference* and denoted by:

$$I_{ij} = I_{r_j}(s_i) = P_{r_j}(s_i) = \frac{P(s_i)}{d_{ij}^{\alpha}}.$$

Note that we do not want to incorporate any near- or mid-field effects. To this end we assume that the setting is normalized, in the sense that the minimum distance between any sender–receiver pair is one, i.e.:

$$d(s_i, r_j) \geq 1, \quad \forall s_i, r_j \mid \ell_i, \ell_j \in L.$$

In other words, without loss of generality, we assume that the input instance L is always *normalized*.

We denote by $\mathcal{S}_t = \{\ell_1, \ldots, \ell_m\}$ the set of concurrently scheduled links in time slot t. As in Ref. [38], we assume that transmissions are slotted into *synchronized time slots* of equal length. In each time slot t, a node can either transmit or remain silent.

The *total interference* $I_{r_i}(\mathcal{S}_t)$ (sometimes also referred to as $I_{\ell_i}(\mathcal{S}_t)$, or simply as I_{r_i} or I_{ℓ_i}) experienced by a receiver r_i is the sum of the interference power values created by all nodes in the network transmitting simultaneously in time slot t (except the intending sender s_i), that is,

$$\begin{aligned} I_{r_i} &= I_{r_i}(\mathcal{S}_t) \\ &- \sum_{\substack{\ell_j \in \mathcal{S}_t, \\ \ell_j \neq \ell_i}} I_{r_i}(s_j). \end{aligned}$$

Let N denote the *ambient noise* power level. We define the *signal-to-interference-plus-noise ratio* of a link ℓ_i, transmitting in time slot t as:

$$\begin{aligned} SINR_{\ell_i} &= SINR_{r_i}(\mathcal{S}_t) \\ &= \frac{P_{ii}}{I_{r_i} + N} \\ &= \frac{\frac{P(s_i)}{d_{ii}^{\alpha}}}{\sum_{\substack{\ell_j \in \mathcal{S}_t, \\ \ell_j \neq \ell_i}} \frac{P(s_j)}{d_{ji}^{\alpha}} + N}. \end{aligned}$$

Finally, let $\beta \geq 1$ denote a hardware-dependent minimum SINR threshold required for a successful message reception. A *successful transmission* between a sender s_i and a receiver r_i in time slot t occurs if and only if:

$$SINR_{\ell_i}(\mathcal{S}_t) \geq \beta. \tag{2.1}$$

We say that a *schedule* $\mathcal{S} = \{\mathcal{S}_1, \ldots, \mathcal{S}_T\}$ is feasible, or *correct*, if the set of communication requests in each time slot is feasible, i.e., if the following condition holds:

$$SINR_{\ell_i}(\mathcal{S}_t) \geq \beta, \quad \forall \ell_i \in \mathcal{S}_t, \quad \forall t \in \{0, \ldots, T - 1\}.$$

The SINR threshold is sometimes referred to as the SIR threshold, when the ambient noise is neglected, or assumed to be zero. It can also be referred to as S/N, in case just the ambient noise is taken into account, i.e., the interference from concurrently scheduled transmissions is assumed to be just noise.

Let us introduce the notion of *link length diversity* $g(L)$, namely the number of magnitudes of distances between senders and receivers in the network.

Definition 2.1. The *link length diversity* $g(L)$ of a set $L = \{\ell_i, \ldots, \ell_n\}$ of communication requests is defined as:

$$g(L) := |\{m | \exists \ell_i, \ell_j \in L : \lfloor \log(d_{ii}/d_{jj}) \rfloor = m\}|. \tag{2.2}$$

A related measure denotes the *link length ratio* between the longest and shortest link.

Definition 2.2. The *link length ratio* Δ of a set $L = \{\ell_i, \ldots, \ell_n\}$ of communication requests is defined as:

$$\Delta := \frac{\max_{l_i \in L} d_{s_i, r_i}}{\min_{l_j \in L} d_{s_j, r_j}}. \tag{2.3}$$

A similar measure denotes the *aspect ratio* between the longest and the shortest distance between any two nodes.[1]

---

[1] The nomenclature between different authors varies. Note, however, that the link length ration and the aspect ratio can differ arbitrarily for the same scenario, e.g., when all links are of length 1, yet the maximum distance d_{max} between two senders is arbitrarily large. For such scenarios, the link length ratio is 1, the aspect ratio is $d_{\max}$.

Definition 2.3. The *aspect ratio* Λ of a set $L = \{\ell_i, \ldots, \ell_n\}$ of communication requests is defined as:

$$\Lambda := \frac{\max_{v_i, v_j \in V} d_{v_i, v_j}}{\min_{v_i, v_j \in V} d_{v_i, v_j}}, \tag{2.4}$$

where the set $V = \{s_1, s_2, \ldots, s_n, r_1, r_2, \ldots, r_n\}$ contains all nodes of L.

We can partition the links into link length classes C_k. A link ℓ_i belongs to class C_k if $2^k \leq d_{ii} < 2^{k+1}$. To obtain the link diversity, we remove the empty classes and count the number of remaining link sets. Observe that the inequality $g(L) \leq \log \Lambda$ holds since there can be at most $\log \Lambda$ link length classes.

2.1.1 Generalized Physical Model

In practice, the received signal power may deviate from the above theoretical bound for various reasons, such as not perfectly omni-directional antennas, shadowing, reflection, or diffraction caused by obstacles. In order to account for some of these aspects, Moscibroda et al. [68] define and study the following generalization of the physical model. In this model, given a parameter θ, the received signal power (as well as the interference caused by simultaneously transmitting nodes) can deviate from the theoretically received power by a factor of at most θ.

Formally, if $P_{r_i}(s_i)$ is the power level received by r_i from a signal transmitted by s_i, the generalized physical model states that $P_{r_i}(s_i)$ is in the range

$$\frac{1}{\theta} \cdot \frac{P(s_i)}{d_{ii}^\alpha} \leq P_{r_i}(s_i) \leq \theta \cdot \frac{P(s_i)}{d_{ii}^\alpha}. \tag{2.5}$$

Note that this model leaves open the exact received signal power. Therefore, algorithms designed to work in this generalized physical model must be robust enough to cope with arbitrary deviations within the stated bounds. Clearly, for $\theta = 1$, the generalized physical model is equivalent to the standard physical interference model.

In wireless communications, the physical model is only considered a base model. On top of pure distance path-loss, stochastic fading modules are added, i.e., Gaussian random variables that model the wireless reality more accurately. Well-known examples are Rayleigh or Rician fading, or dispersive fading models. By incorporating stochastic fading models our simple arguments become much more complicated, in fact, even the problem statements become significantly more complex. Instead of asking simple (deterministic) scheduling questions, we would have to tackle probabilistic questions in this case. We believe that one should study questions regarding probabilistic models as well, however, the basics are better taught using basic models.

2.2 Problem Definitions

In this section we present formal definitions of three variations of the scheduling problem that play a central role in this survey.

Note that instead of assigning links to time slots, one could let them transmit on different channels. The algorithmic problems one has to solve remain the same.

2.2.1 One-Slot Scheduling Problem

The One-Slot Scheduling Problem can be formulated as follows. The input to the problem is a set of links $L = \{\ell_1, \ldots, \ell_n\}$, where each link ℓ_i represents a communication request from a sender s_i to a receiver r_i. The objective of the One-Slot Scheduling Problem is to maximize the number of links scheduled concurrently in one time slot, such that all messages are received successfully. In other words, we attempt to use one slot to its full capacity.

Formally, a set $\mathcal{S} = \{\ell_1, \ldots, \ell_m\} \subseteq L$ is a solution to an instance of the One-Slot Scheduling Problem if the following conditions hold:

$$\mathcal{S} = \operatorname*{argmax}_{\mathcal{S}' \subseteq L} |\mathcal{S}'|,$$

$$SINR_{\ell_i}(\mathcal{S}') \geq \beta, \quad \forall \ell_i \in \mathcal{S}'. \tag{2.6}$$

2.2.2 Weighted One-Slot Scheduling Problem

The Weighted One-Slot Scheduling Problem is a "weighted version" of the One-Slot Scheduling Problem. It can be formulated as follows. The input to the problem is a set of links $L = \{\ell_1, \ldots, \ell_n\}$, where each link ℓ_i is assigned a *weight* $w(\ell_i)$. The weights might represent, for example, the relative priorities of the communication requests, or the revenue values associated with different clients. The objective of the problem is to pick a subset of weighted links, such that the total weight (or value) is maximized and the *SINR* level is at least β at every scheduled receiver.

A set $\mathcal{S} = \{\ell_1, \ldots, \ell_m\} \subseteq L$ is a solution to an instance of the Weighted One-Slot Scheduling Problem if the following conditions hold:

$$\mathcal{S} = \operatorname*{argmax}_{\mathcal{S}' \subseteq L} \sum_{\ell_i \in \mathcal{S}'} w(\ell_i),$$

$$SINR_{\ell_i}(\mathcal{S}') \geq \beta, \quad \forall \ell_i \in \mathcal{S}'. \tag{2.7}$$

2.2.3 Multi-slot Scheduling Problem

As opposed to the one-slot versions of the scheduling problem, where the objective is to use one time slot to its full capacity, the objective of the Multi-slot Scheduling Problem is to schedule all links in as few time slots as possible, guaranteeing that all messages are delivered successfully according to the SINR condition (2.1).

More precisely, let $L = \{\ell_1, \ldots, \ell_n\}$ be the input set of communication requests. A *schedule* is represented by $\mathcal{S} = (\mathcal{S}_1, \mathcal{S}_2, \ldots, \mathcal{S}_{T(\mathcal{S})})$, where $T(\mathcal{S})$ denotes the length of the schedule and $\mathcal{S}_t = \{\ell_1, \ldots, \ell_m\} \subseteq L$ is a subset of links scheduled in time slot t.

A schedule $\mathcal{S}$ is a solution to an instance of the Multi-slot Scheduling Problem if the following conditions hold:

$$\mathcal{S} = \operatorname*{argmin}_{\mathcal{S}' = (\mathcal{S}'_1, \mathcal{S}'_2, \ldots, \mathcal{S}'_{T(\mathcal{S}')})} T(\mathcal{S}'),$$

$$\bigcup_{t=1}^{T(\mathcal{S}')} \mathcal{S}'_t = L,$$

$$SINR_{\ell_i}(\mathcal{S}'_t) \geq \beta, \quad \forall \ell_i \in \mathcal{S}'_t \subseteq \mathcal{S}', \quad t \in \{1, \ldots, T(\mathcal{S}')\}. \tag{2.8}$$

The Multi-slot Scheduling Problem can also be viewed as a coloring problem, in which colors, or labels, have to be assigned to links, such that any subset of links of the same color can transmit successfully according to the SINR constraints.

2.2.4 Scheduling Problems with Power Control

All the problems introduced above can be studied in combination with power control. In this case, a solution to a scheduling problem includes a power assignment for each sender.

2.3 Approximation Algorithms

Many optimization problems are known to be "NP-hard", that is, it is believed that the problems cannot be solved optimally in reasonable time. For NP-hard problems, the best solutions known today need exponential time, as one essentially needs to check all possible solutions. Even for problem instances of moderate size, the fastest computers available need thousands of years.

There are several ways to deal with such problems. One way is to simply propose a heuristic which is then tested by simulation. This is often unsatisfactory, as one cannot be sure that the heuristic will work for all possible problem instances, or whether one just has been lucky with the few that were simulated.

So-called approximation algorithms offer an alternative approach to heuristics. Approximation algorithms find a solution efficiently, no matter how large the problem instance is. We are usually interested in algorithms with polynomial time complexity, i.e., if the size of the problem instance doubles, the execution time of the algorithm may double, or quadruple, or even increase by a factor of 100, but it will not increase exponentially.

However, an approximation algorithm cannot guarantee to find the optimal solution, just a sub-optimal solution. Unlike heuristics, an approximation algorithm always provides a provable solution quality. Ideally, the solution is optimal up to a small constant factor (for instance with at most 1% overhead to the unknown optimal solution).

More formally, an approximation algorithm A achieves an approximation ratio ρ if for all possible inputs I we have

$$cost_A(I) \leq \rho \cdot cost_{OPT}(I),$$

where *cost* is the cost function of the algorithm A and the (unknown) optimal algorithm OPT, respectively. We also say that the algorithm A provides a ρ-approximation and that the problem at hand is ρ-approximable.

In the best case $\rho = 1 + \epsilon$ for an arbitrarily small ϵ; in this case we call the approximation algorithm a polynomialc-time approximation scheme (PTAS). Sometimes, ρ is a constant, independent of the problem size, e.g., $\rho = 2$. Sometimes the best known approximation algorithm for a problem only allows a ρ which is dependent on the input size. If we attempt to minimize the time for scheduling n links, for example, we might have to settle for $\rho = O(\log n)$, i.e., apart from constant factors hidden in the $O()$ notation, the schedule of our approximation algorithm is only guaranteed to be within a logarithmic factor of the input size of the optimal solution. Indeed, there exist optimization problems that are provably impossible to approximate within any constant, or larger function of the input size, if $P \neq NP$.

2.4   Robustness of the Physical Model

The pure geometric quality of interference given in Equation (2.1) is an idealization of true physical characteristics. It assumes, e.g., perfectly isotropic radios, no obstructions, and a constant ambient noise level. That raises the question, why move algorithm analysis from analytically amenable graph-based models to a more realistic model if the latter is not that realistic? Fortunately, the fact that schedule lengths are fairly invariant to signal requirements shows that these concerns are unnecessary.

In this section we discuss robustness properties of the physical model with respect to parameters such as signal strength, or interference tolerance, and spatial dispersion of links. More specifically, in Section 2.4.1, we will show that minor discrepancies in signal strength requirements cause only minor changes in schedule length. Moreover, in Section 2.4.2, we examine the desirable property of link dispersion, and how any schedule can be dispersed at a limited cost.

Throughout the rest of this chapter and the next chapter, we make use of the following definitions.

Definition 2.4. The *relative interference* (RI) of link ℓ_w on link ℓ_v is the increase caused by ℓ_w in the inverse of the SINR at ℓ_v, namely

$$RI_v(w) = \frac{I_{wv}}{P_{vv}}.$$

For convenience, define $RI_v(v) = 0$. Let η_v be a constant that indicates the extent to which the ambient noise approaches the required signal at receiver r_v.[2]

$$\eta_v = \frac{\beta}{1 - \beta \frac{N}{P_{vv}}} = \frac{1}{\frac{1}{\beta} - \frac{N}{P_{vv}}}.$$

The *affectance*[3] of link ℓ_v, caused by a set S of links, is the sum of the relative interferences of the links in S on ℓ_v, scaled by η_v, or

$$a_{\ell_v}(S) = \eta_v \cdot \sum_{\ell_w \in S} RI_v(w). \tag{2.9}$$

For a single link ℓ_w, we use the shorthand $a_{\ell_v}(w) = a_{\ell_v}(\{\ell_w\})$.

Observation 2.5. The affectance function satisfies the following properties for a set S of links:

(1) *Range:* S is SINR-feasible iff, for all $\ell_v \in S$, $a_{\ell_v}(S) \leq 1$.
(2) *Additivity:* $a_{\ell_v}(S) = a_{\ell_v}(S_1) + a_{\ell_v}(S_2)$, whenever (S_1, S_2) is a partition of S.
(3) *Distance bound:* If $P(s_v) = P(s_w)$, then $a_{\ell_v}(\ell_w) = \eta_v \cdot (\frac{d_{vv}}{d_{wv}})^\alpha$, for any pair ℓ_w, ℓ_v in S.

Proof. (1) *Range* $\Rightarrow$: Assume $a_{\ell_v}(S) > 1$. Then

$$a_{\ell_v}(S) = \eta_v \cdot \frac{\sum_{\ell_w \in S} I_{wv}}{P_{vv}} > 1 \Rightarrow \sum_{\ell_w \in S} I_{wv} > \frac{P_{vv}}{\eta_v} = \frac{P_{vv}}{\beta} - N.$$

---

[2] Note that η_v does not have any particular physical meaning and is introduced in order to simplify other definitions.

[3] Affectance is closely related to *affectedness*, defined in Ref. [30], but treats the effect of noise more accurately.

This means that

$$SINR_{\ell_v}(S) = \frac{P_{vv}}{N + \sum_{\ell_w \in S} I_{wv}} < \frac{P_{vv}}{N + \frac{P_{vv}}{\beta} - N} = \beta.$$

(1) *Range* $\Leftarrow$: Assume $a_{\ell_v}(S) \le 1$. Then

$$\sum_{\ell_w \in S} I_{wv} \le \frac{P_{vv}}{\beta} - N \Rightarrow SINR_{\ell_v}(S) = \frac{P_{vv}}{N + \sum_{\ell_w \in S} I_{wv}} \ge \beta.$$

(2) *Additivity*: Since $S_1 \cap S_2 = \emptyset$ and $S_1 \cup S_2 = S$, we have that:

$$a_{\ell_v}(S) = \frac{\eta_v}{P_{vv}} \cdot \sum_{\ell_w \in S} I_{wv}$$

$$= \frac{\eta_v}{P_{vv}} \cdot \left(\sum_{\ell_w \in S_1} I_{wv} + \sum_{\ell_w \in S_2} I_{wv} \right)$$

$$= \frac{\eta_v}{P_{vv}} \cdot \sum_{\ell_w \in S_1} I_{wv} + \frac{\eta_v}{P_{vv}} \cdot \sum_{\ell_w \in S_2} I_{wv}$$

$$= a_{\ell_v}(S_1) + a_{\ell_v}(S_2).$$

(3) *Distance bound*: Since $P(s_v) = P(s_w)$, we have that:

$$a_{\ell_v}(\ell_w) = \eta_v \cdot \frac{I_{wv}}{P_{vv}} = \eta_v \cdot \frac{\frac{P(s_w)}{d_{wv}^\alpha}}{\frac{P(s_v)}{d_{vv}^\alpha}}$$

$$= \eta_v \cdot \left(\frac{d_{vv}}{d_{wv}} \right)^\alpha \cdot \frac{P(s_v)}{P(s_w)} = \eta_v \cdot \left(\frac{d_{vv}}{d_{wv}} \right)^\alpha. \qquad \square$$

Notation 2.6. Furthermore, we list some additional useful notations:

- OPT: an SINR-feasible schedule of optimum size
- χ: minimum number of slots in an SINR-feasible schedule
- S_p: a p-signal schedule, where $a_{\ell_v}(S_p) \le 1/p, \forall \ell_v \in S_p$
- OPT_p: a p-signal schedule of optimum size
- χ_p: minimum number of slots in a p-signal schedule
- $S_{\ell_v}^+$ (or $S_{\ell_v}^-$): set of links in S, which are longer (or shorter) than ℓ_v

2.4.1   Interference Tolerance

In this section we explore how signal requirements (the value of β), or equivalently interference tolerance, affects schedule length. It is not *a priori* obvious that minor discrepancies cause only minor changes in schedule length. However by showing that it is so, we can give the algorithm designer the advantage of being able to compare the algorithms' output with a stricter optimal schedule. This also has implications regarding the robustness of the physical model with respect to perturbations in signal transmissions.

Theorem 2.7 ([42]).   There is a polynomial-time algorithm that takes a p-signal schedule and refines it into a p'-signal schedule, for $p' > p$, increasing the number of slots by a factor of at most $\lceil 2p'/p \rceil^2$.

Proof.   Consider a p-signal schedule $\mathcal{S}_p$ and a slot S_t in $\mathcal{S}_p$. We partition S_t into a sequence $S_{t1}, S_{t2}, \ldots$ of sets (see Figure 2.1). Order the links in S_t in some order, e.g., decreasing order of link length. For each link ℓ_v, assign ℓ_v to the first set S_{ti} for which $a_{\ell_v}(S_{ti} \cap S_{l_v}^+) \le 1/2p'$, i.e., the accumulated affectance on ℓ_v among the previous, longer links in S_{ti} is at most $1/2p'$. Since each link ℓ_v originally had affectance at most $1/p$, then by the additivity of affectance, the number of sets used is at most $\lceil \frac{1/p}{1/2p'} \rceil = \lceil \frac{2p'}{p} \rceil$.

  We then repeat the same approach on each of the sets S_{ti}, processing the links this time in increasing order. The number of sets is again $\lceil \frac{2p'}{p} \rceil$ for each S_{ti}, or $\lceil \frac{2p'}{p} \rceil^2$ in total. In each final slot (set), the affectance on a link by shorter links in the same slot is at most $1/2p'$. In total, then, the affectance on each link is at most $2 \cdot 1/2p' = 1/p'$. $\square$

  Please note that the result only holds in one direction, i.e., we can generally not decrease the slots by lowering p.

  This result however applies to optimal solutions. Let OPT_p, χ_p be defined as in Section 2.6. It is not *a priori* clear that a smooth relationship exists between χ_p and χ, for $p > 1$.

Corollary 2.8.   [42] $\chi_p \le \lceil 2p/\beta \rceil^2 \chi$.

$$
\begin{aligned}
\mathcal{S}_p \;=\; & \{S_1, S_2, S_3, \dots S_t, \dots, S_{\chi_p-1}, S_{\chi_p}\}, \quad a_{\ell_v}(S_t) \leq 1/p, \forall \ell_v \in S_t. \\[4pt]
& \overbrace{\{S_{t1}, \dots, S_{ti}, \dots, S_{tj}\}}^{S_t}, \qquad a_{\ell_v}(S_{ti} \cap S_{l_v}^{+}) \leq 1/2p', \\
& \underrightarrow{\text{first-fit descr. link length}} \qquad \forall \ell_v \in S_{ti}. \\[4pt]
& \qquad\qquad\qquad\qquad a_{\ell_v}(S_{t(i-1)}) > 1/2p' \Rightarrow \\
& \overbrace{\qquad\qquad S_{ti} \qquad\qquad}^{} \qquad j < \lceil 2p'/p \rceil. \\[4pt]
& \{S_{t11}, \dots, S_{til}, \dots, S_{tik}\}, \qquad a_{\ell_v}(S_{til} \cap S_{l_v}^{-}) \leq 1/2p', \\
& \underrightarrow{\text{first-fit incr. link length}} \qquad \forall \ell_v \in S_{til}. \\[4pt]
& \qquad\qquad\qquad\qquad a_{\ell_v}(S_{ti(l-1)}) > 1/2p' \Rightarrow \\
& \qquad\qquad\qquad\qquad k < \lceil 2p'/p \rceil. \\[4pt]
|\mathcal{S}_{p'}| \;\leq\; & |\mathcal{S}_p| \cdot j \cdot k \leq |\mathcal{S}_p| \cdot \lceil 2p'/p \rceil^2.
\end{aligned}
$$

Fig. 2.1 Illustration of Theorem 2.7: p-signal schedule $\mathcal{S}_p$ is transformed into a p'-signal schedule $\mathcal{S}_{p'}$ with at most $|\mathcal{S}_p| \cdot \lceil 2p'/p \rceil^2$ time slots.

This has significant implications. One might question the validity of studying the pure physical model. As we discussed in Section 2.1.1, the received signal power may deviate from the theoretical bound assumed by the "pure" physical model for various reasons. A *generalized physical model*, described in Section 2.1.1, was introduced to allow for such deviations.

Theorem 2.7 implies that scheduling is robust under discrepancies in the physical model, e.g., the results carry over to the generalized physical model. This validates the analytic study of the pure physical model, in spite of its simplifying assumptions.

Corollary 2.9 ([42]). If a scheduling algorithm gives an ρ-approximation in the physical model, it provides an $O(\theta^2 \rho)$-approximation in variations in the physical model with a discrepancy of up to a factor of θ in signal attenuation or ambient noise levels.

This result can be contrasted with the strong $n^{1-\varepsilon}$-approximation hardness of scheduling in an abstract (non-geometric) SINR model that allows for arbitrary distances between nodes [32]. Alternatively, Theorem 2.7 allows us to analyze algorithms under more relaxed situations than the optimal solutions that we compare with.

2.4.2 Spatial Dispersion under Uniform Power Assignment

A popular power assignment scheme used in practical systems is the *uniform power assignment* [38]. In this scheme, all nodes transmit with the same power level.

One desirable property of schedules that employ uniform power assignment is that links in the same slot be spatially well separated, i.e., it is generally desired that links that transmit concurrently should be located relatively far from each other, such that interference is not too high. This however blurs the difference in position between sender and receiver of a link, since it affects distances only by a small constant. The notion of spatial separation should depend on the length of the links themselves, i.e., longer (weaker) links require more spatial separation (i.e., longer distance to concurrent links) than do shorter (stronger) links. Intuitively, we want to measure nearness as a fraction of the lengths of the respective links. Given the affectance measure, it proves to be useful to define it somewhat less restrictively.

Definition 2.10. Link ℓ_w is said to be *q-near* link ℓ_v, if $d_{wv} < q \cdot \eta_v^{1/\alpha} \cdot d_{vv}$. A set of links is *q-dispersed* if no (ordered) pairs of links in the set are *q*-near. A schedule is *q*-dispersed if all the slots are formed by *q*-dispersed sets.

Observation 2.5, item 3, states that link ℓ_w is *q*-near a link ℓ_v iff $a_{\ell_v}(\ell_w) > q^{-\alpha}$. This immediately gives the following lemma.[4]

Lemma 2.11 ([42]). If all nodes transmit with the same power level, fewer than q^α/p senders in a *p*-signal schedule S_p are *q*-near to any given link $\ell_v \in S$.

At a cost of a constant factor, any schedule can be made dispersed.

---

[4] Lemma 2.11 is a strengthening of Lemma 4.2 in Ref. [30].

Lemma 2.12 ([42]). Assume all nodes transmit with the same power level. There is a polynomial-time algorithm that takes an SINR-feasible schedule and refines it into a q-dispersed schedule, increasing the number of slots by a factor of at most $(q + 2)^\alpha$.

Proof. Let S_t be a slot in an SINR-feasible schedule S. We show how to partition S_t into sets $S_{t1}, S_{t2}, \ldots, S_{tj}$ that are q-dispersed, where $j \le (q + 2)^\alpha + 1$ (see Figure 2.2).

Process the links of S_t in increasing order of length, assigning each link ℓ_v "first-fit" to the first set S_{ti} in which the receiver r_v is at least $(q\eta_v^{1/\alpha} + 2) \cdot d_{vv}$ away from any other link. Let ℓ_w be a link previously in S_{ti}, and note that ℓ_w is shorter than ℓ_v. By the selection rule,

$$d_{wv} \ge (q\eta_v^{1/\alpha} + 2) \cdot d_{vv} \ge q\eta_v^{1/\alpha} \cdot d_{vv},$$

$$d_{vw} \ge d_{wv} - d_{ww} - d_{vv} \ge (q\eta_v^{1/\alpha} + 1)d_{vv} - d_{ww} \ge q\eta_v^{1/\alpha}d_{ww}.$$

Since this holds for every pair in the same set, the schedule is q-dispersed. Suppose S_{tj} is the last set used by the algorithm, and

$$
\begin{aligned}
\mathcal{S} \;=\;\; & \{S_1, \ldots S_t, \ldots, S_T\}, && a_{\ell_v}(S_t) \le 1, \forall \ell_v \in S_t. \\[4pt]
& \overbrace{\{S_{t1}, \ldots, S_{ti}, \ldots, S_{tj}\}}^{\textstyle S_l}, && d(s_w, r_v) \ge (q\eta_v^{1/\alpha} + 2)d_{vv}, \\
& \underrightarrow{\text{first-fit incr. link length}} && \forall \ell_w \in S_{ti} \cap S_{l_w}^- \Rightarrow \\
& && d(s_v, r_w) \ge q\eta_v^{1/\alpha}d_{ww}, \\
& && \forall \ell_w \in S_{ti} \cap S_{l_w}^- \Rightarrow \\
& && S_{ti} \text{ is } q\text{-dispersed.} \\
& && d(s_w, r_v) < (q + 2)\eta_v^{1/\alpha}d_{vv}, \\
& && \ell_w \in S_{t(i-1)} \Rightarrow \\
& && j < (q + 2)^\alpha.
\end{aligned}
$$

Fig. 2.2 Illustration of Lemma 2.12: SINR-feasible schedule $\mathcal{S}$ is transformed into a q-dispersed schedule with at most $|\mathcal{S}| \cdot (q + 2)^\alpha$ time slots.

let ℓ_v be a link in it. Then, each S_{ti}, for $i = 1, 2, \ldots, j-1$, contains a link whose sender is closer than $(q\eta_v^{1/\alpha} + 2) \cdot d_{vv} \leq (q+2)\eta_v^{1/\alpha}d_{vv}$ to r_v, i.e., is $(q+2)$-near to ℓ_v. By Lemma 2.11, $j-1 < (q+2)^\alpha$. $\square$

Corollary 2.13. [42] Assume all nodes transmit with the same power level. Let χ^q denote the minimum number of slots in a q-dispersed schedule. $\chi^q \leq (q+2)^\alpha \cdot \chi$.

Intuitively, there is a correlation between low affectance and high dispersion in schedules. The following result makes this connection clearer. The converse is, however, not true, since interference of many far-away links can accumulate.

Lemma 2.14. [42] A p-signal schedule that employs uniform power assignment is also $p^{1/\alpha}$-dispersed.

Proof. Let ℓ_v and ℓ_w be an ordered pair of links in a slot S_t in a p-signal schedule. By definition, $a_{\ell_v}(\ell_w) \leq a_{\ell_v}(S_t) \leq 1/p$. By Observation 2.5, item 3, $d_{wv} \geq p^{1/\alpha}\eta_v^{1/\alpha} \cdot d_{vv}$. Hence, the lemma follows. $\square$

2.5   SINR Reception Diagrams

The SINR diagram of a set of sender nodes partitions the plane into regions or reception zones. A region contains the set of locations where the signal of a certain transmitter can be received successfully. Avin et al. [3]. investigate the shape of the reception zones. They show that the reception zones of all senders are convex for a uniform scheme but not necessarily for non-uniform power assignments. In addition, they prove that the reception zones are fat, i.e., the ratio between the radius of the smallest ball completely containing the zone and the radius of the largest ball completely contained in the zone is bounded by a constant. The study of SINR diagrams helps to understand the properties of the physical model better and may lead to more sophisticated algorithms exploiting the characteristics of wireless networks.

2.6 Outlook

In this section we presented properties of schedules in the physical model, which serve as tools for the algorithm designer. Note that the interference tolerance properties, presented in Section 2.4.1, apply equally to scheduling links of constant or fixed power levels (Section 3), as well as variable power levels (Section 4). The spatial dispersion properties, presented in Section 2.4.2, apply for scenarios that employ uniform power assignment, and are going to be used in the analysis presented in Section 3.

3

Scheduling Without Power Control

This section studies the problem of scheduling links in the physical model under the assumption that all nodes transmit with the same power level, i.e., there is no power control. This kind of power assignment is employed in many practical systems, and is sometimes referred to as *uniform power assignment* [38]. It turns out that even in this simplified setting, the problem presents many challenges. We start the chapter with a discussion on the problem's complexity in Section 3.1, where we present a proof that the problem is NP-hard in the physical model, even in a geometry-restricted setting. The rest of the chapter is dedicated to approximation algorithms for the problem. In Section 3.2 we present the first and rather naive approach to the problem. In Section 3.3 we present the most recent results in the area, which prove that the One-Slot Scheduling Problem can be approximated to within a constant factor and the Multi-slot Scheduling Problem can be approximated to within an $O(\log n)$ factor.

3.1 Complexity in the Physical Model

Many complexity results in wireless networks are derived in a general SINR model, where the gain between two transmissions is defined in an arbitrary way, i.e., without considering the geometry of the problem. We refer to such a model as *"abstract"* signal-to-interference-plus-noise-ratio (or short, $SINR_A$) model. In the $SINR_A$ model, the path-loss between nodes is not constrained by their Euclidean coordinates, but can be set arbitrarily (i.e., triangular inequality must not be preserved when defining the path-loss). We illustrate how a typical hardness proof in the $SINR_A$ model works in Section 3.1.1.

The majority of the results analyzed in this monograph are derived in the *geometric* version of the physical model. In this model, nodes live in space, and the gain (or signal attenuation) between two nodes is determined by the distance between the two nodes.

The physical model makes some simplifying assumptions, such as perfectly isotropic radios, no obstructions, or a constant ambient noise level. On the other hand, $SINR_A$ is not all that realistic either, as it allows arbitrary values in the gain matrix among the participating nodes of a wireless network. In reality, if a node u is close to a node v, which in turn is close to a node w, then u and w will also be close. So the entries in the gain matrix will be constrained by the other entries. Thus, the physical model is too optimistic, whereas $SINR_A$ is too pessimistic. Hence, a real network is positioned somewhere between the $SINR_A$ model and the physical model, i.e., the propagation of the wireless signal is neither as well-behaved as in the geometric physical model, nor is it as unpredictable as in the general model.

When studying algorithms or protocols, upper bounds should be derived for the pessimistic model, as an algorithm for a strictly[1] more pessimistic model will also work for reality. However, also the converse is true: if one is interested in lower bounds (impossibility results or capacity constraints), one must use the optimistic model. A strictly more optimistic model guarantees that results are applicable in practice.

In Section 3.1.2, we prove that the Multi-slot Scheduling Problem is NP-hard in the physical model. Since this model is weaker than reality,

---

[1] Note that models are rarely strictly harder than reality; $SINR_A$ is a typical example, as $SINR_A$ does not include several difficulties of reality, e.g., short-term fading.

this implies that one cannot compute an optimal schedule of wireless requests in practice, unless P = NP.

3.1.1 Complexity of the Multi-slot Scheduling Problem in $SINR_A$

In this section we show that it is NP-hard to approximate the Multi-slot Scheduling Problem in the $SINR_A$ model to within a factor of $n^{1-\varepsilon}$, for any constant $\varepsilon > 0$.

Theorem 3.1 ([30]). There is no $n^{1-\varepsilon}$ factor approximation algorithm for the Multi-slot Scheduling Problem in the $SINR_A$ model, for any constant $\varepsilon > 0$, assuming P $\neq$ NP.

Proof. We will prove the result by presenting a *gap-preserving* reduction from the graph coloring problem. In Ref. [89] it was shown that it is NP-hard to approximate the graph coloring problem to within $n^{1-\varepsilon}$ for all $\varepsilon > 0$.

Consider an instance π_C of the graph coloring problem defined for an undirected graph $G = (V, E)$ on n vertices. We construct (in polynomial time) an instance π_S of scheduling, such that:

$$OPT(\pi_C) \leq k \quad \Rightarrow \quad OPT(\pi_S) \leq k, \tag{3.1}$$

$$OPT(\pi_C) > n^{1-\varepsilon}k \quad \Rightarrow \quad OPT(\pi_S) > n^{1-\varepsilon}k. \tag{3.2}$$

For each $v \in V$, we add a link $l_v = (r_v, s_v)$. The physical model parameters are set to $\beta = 1, N = 0$, and the $n \times n$ path-loss matrix A is defined as follows:

- $(v, w) \in E \quad \Rightarrow \quad A_{wv} = A_{vw} = 1,$
- $(v, w) \notin E \quad \Rightarrow \quad A_{wv} = A_{vw} = n,$
- $v = w \quad \Rightarrow \quad A_{wv} = A_{vw} = 1.$

To see that Equation (3.1) holds, assume that we can color π_C with k or less colors. We claim that links associated to nodes with the same color (let's call each such subset $V(c_i), 1 \leq i \leq k$) can be scheduled concurrently, giving a schedule of length k (or less). Since nodes colored with the same color are not adjacent, the SINR at each link l_v can be lower bounded by

$$SINR_{l_v}(V(c_i)) = \frac{\frac{P}{1}}{\sum_{\substack{w \in V(c_i), \\ w \neq v}} \frac{P}{n}}$$

$$\geq \frac{n}{n-1}$$

$$> 1$$

$$= \beta, \quad \forall l_v, \quad v \in V(c_i), \quad i \in \{1, \ldots, k\}.$$

To see that Equation (3.2) holds, assume we cannot color π_C with $\leq n^{1-\varepsilon}k$ colors. We have to show that π_S cannot be scheduled in $n^{1-\varepsilon}k$ time slots or less. Assume that we could, and consider a schedule of size $n^{1-\varepsilon}k$. Since any coloring of this size must have a violation (an edge to a node x of the same color) at at least one node $v \in V$. If s is the color of v, i.e., $v \in V(c_s)$, the SINR at the link l_v associated with this node is:

$$SINR_{l_v}(V(c_s)) \leq \frac{\frac{P}{1}}{\frac{P}{1} + \sum_{\substack{w \in V(c_s), \\ w \neq v, w \neq x}} \frac{P}{n}}$$

$$< 1$$

$$= \beta, \quad l_v \mid v, x \in V(c_s), \quad s \in \{1, \ldots, k\}.$$

This shows that any schedule of size $n^{1-\varepsilon}k$ or less will have at least one violated node, given the necessary contradiction. $\qquad\square$

3.1.2   Complexity of the Multi-slot Scheduling Problem in the Physical Model

Proving a problem to be NP-hard implies that there exists no polynomial time algorithm for determining an optimal schedule, unless $P = NP$. It is widely believed that an NP-hard computational problem is not tractable efficiently.

We proceed by first showing that the decision version of the Multi-slot Scheduling Problem under uniform power assignment scheme is NP-hard by providing a polynomial time reduction from the Partition

Problem, an NP-complete special case of the well-known Subset Sum problem [27]. If the solution to an instance of the Multi-slot Scheduling Problem implies a solution to any instance of the Partition Problem, Multi-slot Scheduling Problem must be at least as hard as Partition.

Theorem 3.2 ([32]). The Multi-slot Scheduling Problem is NP-hard.

Proof. We show that the Partition Problem is reducible to the Multi-slot Scheduling Problem in polynomial time. The Partition Problem (proved to be NP-complete by Karp in his seminal work [48]) can be formulated as follows: given a set $\mathcal{I}$ of integers, is it possible to divide this set into two subsets $\mathcal{I}_1$ and $\mathcal{I}_2$, such that the sums of the numbers in each subset are equal? The subsets $\mathcal{I}_1$ and $\mathcal{I}_2$ must form a partition in the sense that they are disjoint and they cover $\mathcal{I}$.

Partition Problem: Find $\mathcal{I}_1, \mathcal{I}_2 \subset \mathcal{I} = \{i_1, \ldots, i_n\}$ s.t.:

$$\mathcal{I}_1 \cap \mathcal{I}_2 = \emptyset,$$

$$\mathcal{I}_1 \cup \mathcal{I}_2 = \mathcal{I}, \quad \text{and}$$

$$\sum_{i_j \in \mathcal{I}_1} i_j = \sum_{i_j \in \mathcal{I}_2} i_j = \frac{1}{2} \sum_{i_j \in \mathcal{I}} i_j.$$

The proof proceeds as follows. First, we define a many-to-one reduction from any instance of the Partition Problem to a geometric instance of the Multi-slot Scheduling Problem. Then, we argue that the instance of the Multi-slot Scheduling Problem cannot be scheduled in $T \leq 1$ time slots, but can be scheduled in $1 < T \leq 2$ time slots if and only if the instance of the Partition Problem is solved.

Let us look at a set $\mathcal{I} = \{i_1, \ldots, i_n\}$ of integers, where the elements of $\mathcal{I}$ add up to σ,

$$\sum_{j=1}^{n} i_j = \sigma.$$

Without loss of generality, we can assume all elements to be distinct and positive.[2] We construct the following Multi-slot Scheduling

---

[2] Note that the assumption that the integers in the Partition instance are distinct is not essential for the reduction to work, and we make it merely for the sake of simplicity.

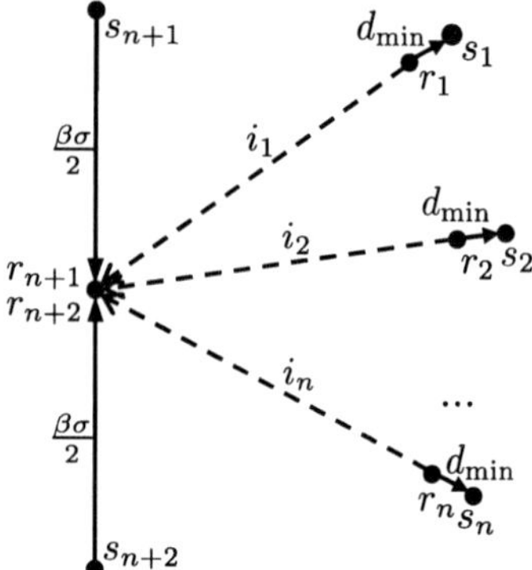

Fig. 3.1 Reduction from Partition: link l_{n+1} (or l_{n+2}) can be scheduled if and only if the interference caused by simultaneously scheduled links $s_j, j \in \{1\cdots n\}$ is less or equal to $\sigma/2$.

Problem instance with $n + 2$ links $L = \{l_1, \ldots, l_{n+2}\}$ (see Figure 3.1). We refer to the sender node belonging to l_j as s_j and the receiver node r_j. We assign each of these nodes a position (X,Y) in the plane. For each integer i_j in I we set the x-axis coordinate of s_j to $(P/i_j)^{1/\alpha}$,

$$pos(s_j) = \left(\left(\frac{P}{i_j} \right)^{\frac{1}{\alpha}}, 0 \right) \quad \forall 1 \leq j \leq n.$$

Next, we designate for every $r_i, 1 \leq i \leq n$ its position to be at distance $d_{\min}$ to its sender s_i, where

$$d_{\min} = P^{\frac{1}{\alpha}} \cdot \frac{\left(\frac{1}{(i_{\max}-1)^{1/\alpha}} - \frac{1}{i_{\max}^{1/\alpha}} \right)}{\left(1 + (n\beta)^{\frac{1}{\alpha}} \right)} 3, \tag{3.3}$$

and $i_{\max}$ is the maximal value of the integers in set $\mathcal{I}$. Thus

$$pos(r_i) = pos(s_i) + (d_{\min}, 0).$$

---

[3] Note that this implies that some sender–receiver distances are less than one and the received power $P_{r_i} = P/d^{\alpha}_{\min}$ can be larger than the transmitting power P. As has been stated in Section 2.1, to overcome this issue, we assume that the problem instance is normalized such that the minimum distance between any sender–receiver pair is at least one. The power level P is normalized accordingly, such that it is high enough for the longest link in the input set to transmit successfully in the presence of ambient noise. For the sake of simplicity, we do not change the notation to reflect this normalization. Power P therefore denotes an already normalized constant. Note that the exact value of P does not affect the reduction, since P is still uniform (fixed and equal for all nodes) and can be determined for any instance of Partition.

Finally, we place r_{n+1} and r_{n+2} at the center $(0,0)$ and their senders s_{n+1}, s_{n+2} perpendicular to the x-axis, at distance $(2P/\beta\sigma)^{1/\alpha}$, i.e.,

$$pos(r_{n+1}) = pos(r_{n+2}) = (0,0),$$

$$pos(s_{n+1}) = \left(0, \left(\frac{2P}{\beta \cdot \sigma}\right)^{\frac{1}{\alpha}}\right),$$

$$pos(s_{n+2}) = \left(0, -\left(\frac{2P}{\beta \cdot \sigma}\right)^{\frac{1}{\alpha}}\right).$$

Having defined the geometric instance of the Multi-slot Scheduling Problem for any instance of the Partition Problem, we proceed by showing that in order to find a schedule of length $1 < T \leq 2$, a solution to the Partition Problem is required. Clearly, it is not possible to schedule all links in one slot, since the receivers r_{n+1} and r_{n+2} are at the same position and we assume $\beta \geq 1$.

In order to transmit successfully, the *SINR* constraint at the intended receiver has to be satisfied. In the following lemma we prove that the receivers $r_1, \ldots, r_n$ are close enough to their respective senders to guarantee successful transmission, regardless of the number of other links scheduled simultaneously.

Lemma 3.3 ([32]). Let $L_i = \{l_j \mid 1 \leq j \leq n + 1 \text{ and } i \neq j\}$. It holds for all $i \leq n$ that the *SINR* exceeds β when the link l_i is scheduled concurrently with the set L_i,

$$SINR_{r_i}(L_i) = \frac{\frac{P}{d_{ii}^{\alpha}}}{\sum_{l_j \in L_i} \frac{P}{d_{ji}^{\alpha}}} > \beta.$$

Proof. We do not consider l_{n+2}, since the interference at r_{n+1} and r_{n+2} is the same and they can never be scheduled simultaneously. Hence the result carries over to l_{n+2}.

Since the positions of the sender nodes $s_1, \ldots, s_n$ depend on the values of $i_1, \ldots, i_n$, we can determine the minimum distance between two sender nodes s_j, s_k.

$$d(s_j, s_k) = |d(s_j, r_{n+1}) - d(s_k, r_{n+1})|$$

$$= \left| \left(\frac{P}{i_j} \right)^{\frac{1}{\alpha}} - \left(\frac{P}{i_k} \right)^{\frac{1}{\alpha}} \right|$$

$$\geq P^{\frac{1}{\alpha}} \left(\frac{1}{(i_{\max} - 1)^{1/\alpha}} - \frac{1}{i_{\max}^{1/\alpha}} \right). \tag{3.4}$$

Thus, one can deduce that the sender s_j closest to $r_i, i \neq j$ is located at least at distance $d(s_j, s_i) - d_{\min}$ from r_i ($d_{\min}$ is defined in Equation (3.3)). All the other sender nodes (including s_{n+1}) are farther away. This suffices to show a lower bound for $SINR_{r_i}(L_i)$.

$$SINR_{r_i}(L_i) > \frac{\frac{1}{d_{\min}^{\alpha}}}{\frac{n}{(d(s_j, s_i) - d_{\min})^{\alpha}}}$$

$$\geq \frac{1}{n}((1 + (n\beta)^{\frac{1}{\alpha}}) - 1)^{\alpha}$$

$$= \beta. \tag{3.5}$$

$\square$

Having proved that successful transmission is guaranteed for links $l_1, \ldots l_n$, no matter how many other links are scheduled concurrently, we now return to the proof of Lemma 3.2.

We claim that there exists a solution to the Partition Problem if and only if there exists a 2-slot schedule for L. For the first part of the claim, assume we know two subsets $\mathcal{I}_1, \mathcal{I}_2 \subset \mathcal{I}$, whose elements sum up to $\sigma/2$. To construct a 2-slot schedule, $\forall i_j \in \mathcal{I}_1$, we assign the link l_j to the first time slot, along with l_{n+1}, and assign the remaining links to the second time slot. Due to Lemma 3.3 we can focus our analysis on the receivers r_{n+1} and r_{n+2}. The situation is the same for both receivers, so it suffices to examine r_{n+1}. The signal power r_{n+1} receives from its sender node s_{n+1} is:

$$P_{r_{n+1}}(s_{n+1}) = \frac{P}{\left(\left(\frac{2P}{\beta\sigma} \right)^{\frac{1}{\alpha}} \right)^{\alpha}} = \frac{\beta\sigma}{2}.$$

The interference r_{n+1} experiences from each sender s_j is:

$$I_{r_{n+1}}(s_j) = \frac{P}{\left(\left(\frac{P}{i_j}\right)^{\frac{1}{\alpha}}\right)^{\alpha}} = i_j,$$

which results in total interference of

$$I_{r_{n+1}} = \sum_{i_j \in \mathcal{I}_1} i_j = \frac{\sigma}{2}.$$

This allows to lower bound the *SINR* at r_{n+1}:

$$SINR_{r_{n+1}} \geq \frac{P_{r_{n+1}}(s_{n+1})}{I_{r_{n+1}}} = \frac{\beta\sigma/2}{\sigma/2} = \beta,$$

which, in combination with Lemma 3.3, proves that our schedule guarantees successful transmission for all links.

For the second part of the claim, we need to show that if no solution to the Partition Problem exists, we cannot find a 2-slot schedule for L. No solution to the Partition Problem implies that for every partition of $\mathcal{I}$ into two subsets, the sum of one set is greater than $\sigma/2$. Assume we could still find a schedule with only two slots. Since the receivers r_{n+1} and r_{n+2} are at the same position, they have to be assigned to different slots to permit a successful transmission. Because we have to split $L \setminus \{l_{n+1}, l_{n+2}\}$ into two sets and the received power from $s_j, j = 1,\ldots,n$ at $(0,0)$ is i_j, we end up with a total interference at $(0,0)$ greater than $\sigma/2$ for one slot, which prevents the correct reception of the signal from s_{n+1} or s_{n+2}. $\qquad\square$

It can be shown that the decision version of the Weighted One-Slot Scheduling Problem with uniform power assignment scheme is also NP-hard in the physical model. The proof is similar in spirit to the one presented above, and the reduction is done from the Knapsack Problem. Please refer to Ref. [32] for more details.

3.1.3   Outlook

In this section we have established that the Multi-slot Scheduling Problem (and its weighted version) is NP-hard in the physical model. In order to prove that the problem at hand is also NP-*complete*, we have to prove that it is in the complexity class NP. For some operations on integers it is not yet clear whether they can be computed efficiently by a Turing machine. E.g., it is not known how a sum of square roots of integers can be compared quickly to an integer [69]. Since the physical model requires the computation of roots of integers, we do not know whether scheduling and related problems are in NP. If we assume the Real RAM (Random Access Machine) model (often used in computational geometry), all the necessary computations can be implemented efficiently.

In the rest of this chapter, we will turn our attention to the design of efficient approximation algorithms. In Sections 3.2 and 3.3, we propose scheduling algorithms that compute feasible solutions in the physical model in polynomial time with worst-case approximation guarantees for arbitrary network topologies.

3.2 Diversity Scheduling

Solving problems in the physical model is difficult, as is documented by the vast amount of literature with heuristics on this subject [6, 10, 11, 12, 26, 34, 37, 45, 63, 64, 65, 66, 68, 77, 83].

In this section we present an algorithm for the Multi-slot Scheduling Problem. This algorithm represents an initial effort to solve this problem in the physical model, and an algorithm with improved approximation guarantee is going to be presented in Section 3.3.

Before describing the algorithm, let us recall the notion of *link length diversity* $g(L)$, defined in Section 2.1, and denoting the number of nonempty length classes of the set of links to be scheduled. Note that, in realistic scenarios, the link length diversity can usually be regarded as a constant. In theory, however, it can be as large as n, the number of links in the network.

The algorithm presented in this section consists of two steps: first, the problem instance is partitioned into disjoint link length classes; then, a feasible schedule is constructed for each length class using a greedy strategy.

3.2.1 Algorithm with Approximation Factor $O(g(L))$ for the Multi-slot Scheduling Problem

The algorithm presented in this section is inspired by the heuristic proposed by Moscibroda et al. [66], which schedules a strongly connected set of links in the physical model using linear power assignment. Although similar in spirit, the algorithm in Ref. [66] is not designed to schedule an arbitrary set of links and does not provide an approximation guarantee for the obtained solution.

Algorithm 1 Approximation Algorithm for the Multi-slot Scheduling Problem [32]

1: **input:** Set of links $L = \{l_1, \ldots, l_n\}$;
2: **output:** Schedule $\mathcal{S} = \{\mathcal{S}_1, \ldots, \mathcal{S}_T\}$;
3: Let $G = \{G_0, \ldots, G_{\lfloor \log(\max d_{ii}) \rfloor}\}$ such that G_k is the set of links l_i of length $2^k \le d_{ii} < 2^{k+1}$;
4: Set μ according to (3.6);
5: $t := 0$;
6: **for all** $G_k \ne \emptyset$ **do**
7:     Partition the plane into squares of width $\mu \cdot 2^k$;
8:     4-color the squares such that no two adjacent squares have the same color;
9:     **for** $j = 1$ **to** 4 **do**
10:         **repeat**
11:             **for all** squares A_j^k of width $\mu \cdot 2^k$ and color j **do**
12:                 Pick one not yet scheduled link $l_i \in G_k$ with receiver r_i in A_j^k; (if there is any such l_i left unscheduled)
13:                 $L_j^k := L_j^k \cup l_i$; (schedule l_i in time-slot t)
14:             **end for**
15:             $t := t + 1$;
16:             $\mathcal{S}_t := L_j^k$;
17:         **until** all links with receivers in any square A_j^k have been scheduled
18:     **end for**
19: **end for**
20: **return** $\mathcal{S}$;

The algorithm (for a description in pseudo-code see Algorithm 3.2.1) starts by partitioning the input set of links L into $\lceil \log (\max d_{ii}) \rceil$ (where $\max d_{ii}$ is the length of the longest link $l_i \in L$) possibly empty length classes. Each length class G_k is scheduled separately. First, the plane is partitioned into square grid cells of side $\mu \cdot 2^k$, where μ is defined as follows:

$$\mu = 4 \left(8\beta \cdot \frac{(\alpha - 1)}{(\alpha - 2)} \right)^{\frac{1}{\alpha}}, \tag{3.6}$$

and then the grid cells are colored regularly with four colors (see Figure 3.2). Links whose receivers belong to different squares of the same color are scheduled simultaneously. Note that the inner *repeat* loop (lines 10–17) constructs a schedule of length $\Delta_j^k = \max_{A_j^k \in G_k} (|A_j^k|)$, which is the maximum number of links in length class G_k, whose receivers are in the same grid cell of color j. Given that there are four colors and $g(L)$ nonempty length classes, all links are scheduled in $4\Delta g(L)$ time slots, where $\Delta = \max_{A_j^k \in \{G_0,\dots,G_{\lfloor \log (\max d_{ii}) \rfloor}\}} (|A_j^k|)$.[4]

We show now that the schedule obtained by Algorithm 3.2.1 is correct, by proving in Theorem 3.4 that all links can be scheduled successfully in their respective time slot.

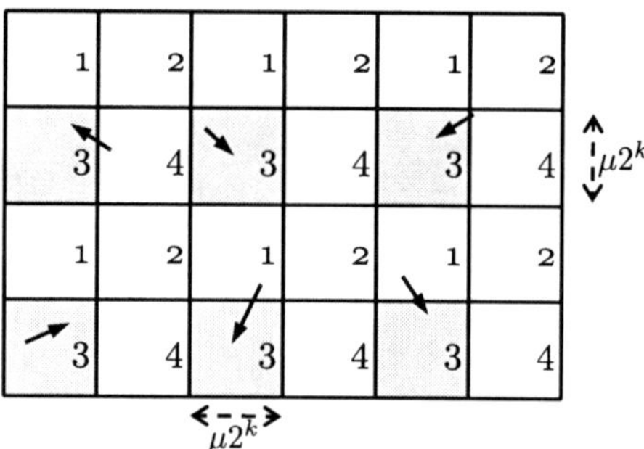

Fig. 3.2 In line 11 of Algorithm 1, the algorithm picks all squares colored with color j. The example shows an inner loop iteration for length class G_k and $j = 3$. The algorithm schedules one unscheduled link from each selected square (if there exists one).

---

[4] Here we overload the term A_j^k to denote the set of receivers $r_i \mid l_i \in G_k$, located inside the grid cell A_j^k; and the term G_k to denote the grid comprising cells of width $\mu \cdot 2^k$.

Theorem 3.4 ([32]). Consider an arbitrary set of links L to be scheduled. For every time slot t, the set $\mathcal{S}_t$ of links output by Algorithm 3.2.1 is scheduled successfully, i.e., the *SINR* at every intended receiver is larger than β.

Proof. We demonstrate that all transmissions scheduled in a time slot t are received successfully by the intended receivers, i.e., their *SINR* is sufficiently high.

Without loss of generality, let us examine links in a length class G_k. Every link $l_i \in G_k$ satisfies $d_{ii} < 2^{k+1}$, thus the perceived power at r_i from s_i is at least

$$P_{r_i}(s_i) \geq \frac{P}{2^{\alpha(k+1)}}. \tag{3.7}$$

Since Algorithm 3.2.1 schedules at most one link in each cell with the same color concurrently, the closest eight senders s_j scheduled in the same time slot must be at least at distance $d(r_i, s_j) \geq \mu 2^k - 2^{k+1} = 2^k(\mu - 2)$ to r_i (see Figure 3.2). Consequently, the sum of their interference experienced by r_i is less than

$$\sum_{j=1}^{8} P_{r_i}(s_j) \leq \frac{8P}{(2^k(\mu - 2))^\alpha}.$$

In the next step, we consider the (at most) 16 senders s_j at distance $3\mu 2^k - 2^{k+1} \leq d(r_i, s_j) \leq 5\mu 2^k - 2^{k+1}$. They contribute a total interference of

$$\sum_{j=9}^{25} P_{r_i}(s_j) \leq \frac{16P}{(2^k(3\mu - 2))^\alpha}.$$

We continue aggregating the interference from nodes s_j at distance range

$$(2l - 1)\mu 2^k - 2^{k+1} \leq d(r_i, s_j) < (2l + 1)\mu 2^k - 2^{k+1},$$

$\forall l = 1, 2, \ldots$. Since at most $8l$ links are picked in each interval, the interference caused by them is at most

$$\sum_{\substack{d(r_i,s_j) \geq \\ (2l-1)\mu 2^k - 2^{k+1}}}^{\substack{d(r_i,s_j) < \\ (2l+1)\mu 2^k - 2^{k+1}}} P_{r_i}(s_j) \leq \frac{8P \cdot l}{(2^k((2l-1)\mu - 2))^\alpha}.$$

Thus, the total interference at a scheduled receiver r_i can be upper bounded by

$$\begin{aligned}
I_{r_i} &\leq \sum_{l=1}^{\infty} \frac{8P \cdot l}{(2^k((2l-1)\mu - 2))^\alpha} \\
&\leq \frac{8P}{2^{k\alpha}} \sum_{l=1}^{\infty} \frac{l}{(\frac{1}{2}(2l-1)\mu)^\alpha} \qquad\qquad (3.8) \\
&\leq \frac{8P}{2^{(k-1)\alpha}\mu^\alpha} \sum_{l=1}^{\infty} \frac{l}{(2l-1)^\alpha} \\
&\leq \frac{8P}{2^{(k-1)\alpha}\mu^\alpha} \sum_{l=1}^{\infty} \frac{1}{l^{\alpha-1}} \\
&\leq \frac{8P}{2^{(k-1)\alpha}\mu^\alpha} \frac{(\alpha-1)}{(\alpha-2)}, \qquad\qquad (3.9)
\end{aligned}$$

where Equation (3.8) follows because $x - 2 > x/2$, $\forall x > 4$ and $\mu > 4$, given that $\beta \geq 1$ and $\alpha > 2$; and Equation (3.9) follows from a bound on Riemann's zeta function. Using Equations (3.7), (3.9), and plugging in the value of μ, defined in Equation (3.6), the *SINR* at receiver r_i can be lower bounded by

$$\begin{aligned}
SINR_{r_i} &= \frac{P_{r_i}(s_i)}{I_{r_i}} \\
&> \frac{\frac{P}{2^{\alpha(k+1)}}}{\frac{8P}{2^{(k-1)\alpha}\mu^\alpha} \frac{(\alpha-1)}{(\alpha-2)}} \\
&= \frac{\mu^\alpha}{4^\alpha \cdot 8 \cdot \frac{(\alpha-1)}{(\alpha-2)}} \\
&= \beta, \qquad\qquad\qquad\qquad\qquad \square
\end{aligned}$$

Now we turn our attention to the efficiency of Algorithm 3.2.1. In particular, in Theorem 3.5 we bound its approximation ratio.

Theorem 3.5 ([32]). The approximation ratio of Algorithm 3.2.1 is $O(g(L))$, where $g(L)$ is the *length diversity* of the input, defined in Definition 2.1.

Proof. The proof relies on the choice of a so-called *critical grid cell*[5]

$$A_{\max}^k = \underset{A_j^k \in \{G_0,\ldots,G_{\lfloor \log(\max d_{ii}) \rfloor}\}}{\mathrm{argmax}} |A_j^k|, \tag{3.10}$$

i.e., we choose the cell with the highest density $\Delta = |A_{\max}^k|$ over all $g(L)$ generated grids (see Figure 3.3). Note that Δ is the number of links l_i whose receiver is located in cell $A_{\max}^k$ and whose length class is G_k, i.e., $2^k \leq d_{ii} < 2^{k+1}$. We proceed by showing that an optimum algorithm OPT can schedule all Δ in at least $T_{OPT} \geq \lceil \Delta/q \rceil$ time slots, where q is a constant dependent on parameters α and β (μ is defined in Equation (3.6)):

$$q = \frac{(2(\sqrt{2}\mu + 1))^\alpha}{\beta}. \tag{3.11}$$

Assume, by contradiction, that OPT schedules all links in less than T_{OPT} time slots. Therefore, there must exist a time slot t',

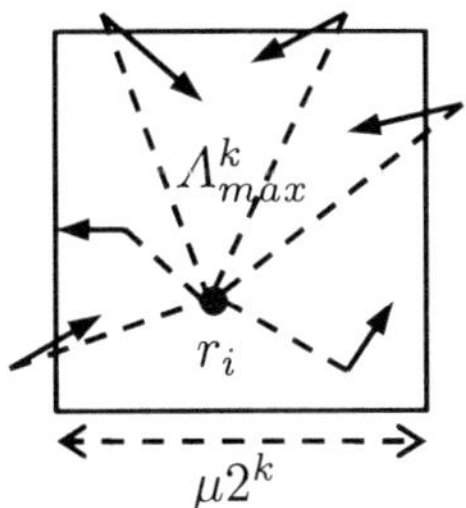

Fig. 3.3 Lower Bound: an optimum algorithm could schedule at most q links with receivers in A_{max}^k in length class G_k in a single time slot.

---

[5] Here we overload the term A_j^k to denote the set of receivers $r_i \mid l_i \in G_k$, located inside the grid cell A_j^k; and the term G_k to denote the grid comprising cells of width $\mu \cdot 2^k$.

$1 \leq t' \leq T_{OPT}$, such that more than q links in $A^k_{\max}$ are scheduled simultaneously. We pick one of the scheduled links $l_i, r_i \in A^k_{\max}$ in time slot t' and calculate the resulting *SINR* level at r_i:

$$SINR_{r_i \in A^k_{\max}} \leq \frac{\frac{P}{d^\alpha_{ii}}}{P \cdot \sum_{j=0}^q d(s_j, r_i)^{-\alpha}}$$

$$< \frac{\frac{P}{2^{k\alpha}}}{P \cdot q \cdot (2\sqrt{2}\mu 2^k + 2^{k+1})^{-\alpha}} \qquad (3.12)$$

$$= \beta, \qquad (3.13)$$

where Equation (3.12) follows from the fact that $d_{ii} \geq 2^k$, $d_{jj} < 2^{k+1}$ and $d(r_i, r_j) \leq 2\sqrt{2}\mu 2^k$; and Equation (3.13) follows from Definition ((3.11)) of q.

Hence, to schedule all links in the *critical* square $A^k_{\max}$, OPT needs time

$$T_{OPT} \geq \left\lceil \frac{\Delta}{q} \right\rceil. \qquad (3.14)$$

On the other hand, Algorithm 3.2.1 schedules all links in L in time

$$T_{ALG\,3.2.1} \leq 4 \cdot \Delta \cdot g(L). \qquad (3.15)$$

The approximation ratio follows from Equations (3.14) and (3.15):

$$\frac{T_{ALG\,3.2.1}}{T_{OPT}} \leq 4q \cdot g(L)$$

$$= O(g(L)). \qquad (3.16)$$

$\square$

Algorithm 3.2.1 can be adapted to solve the Weighted One-Slot Scheduling Problem with the same asymptotic approximation ratio. Please refer to Ref. [32] for more details.

3.2.2   Outlook

The approximation ratio of the algorithm discussed in this section is $O(g(L))$. Although this is the first result to provide any approximation guarantee for the Multi-slot Scheduling Problem in the physical model, it leaves a lot of space for improvement, since this guarantee can become extremely bad $(\Omega(n))$, depending on the topology of the network. In other words, it is not better than the guarantees offered by the most naive solutions to the problem.

In the next section we are going to present an improved scheduling algorithm, whose approximation guarantee no longer depends on the diversity $g(L)$ or any other topological characteristic of the network.

3.3 Approximative Scheduling

In this section we will discuss an algorithm that provides a solution to the One-Slot Scheduling Problem with constant approximation guarantee. When applied repeatedly, it provides an $O(\log n)$ approximation for the Multi-slot Scheduling Problem. As opposed to the algorithm presented in Section 3.2, the approximation guarantee of this algorithm is independent of the topology of the network.

3.3.1 Scheduling Algorithm with Constant Approximation Factor for the One-Slot Scheduling Problem

The one-slot scheduling algorithm (for a description in pseudo-code see Algorithm 2) schedules links in non-decreasing order of their length, say $l_1, \ldots, l_n$. For each link l_v, it checks whether it is too much affected by previously scheduled links (those in the output one-slot schedule S) and, if not, adds it to the solution S. "Too much affected" is defined by a threshold $\tau^{-\alpha}$, where $\tau = 2 + c$, and c is a constant defined in Equation (3.18). All the links that do not satisfy the affectance condition are not added to the solution. We will show that this simple algorithm produces a constant approximation to the One-Slot Scheduling Problem.

We start the analysis by showing that the algorithm obtains an SINR-feasible schedule.

3.3.1.1 Correctness

In this section we prove that the solution S output by Algorithm 2 is SINR-feasible, i.e., all links assigned to the same time-slot can be scheduled concurrently without collisions.

Algorithm 2 One-Slot Scheduling Algorithm [42]

1: **input:** Set of links $L = \{l_1, \ldots, l_n\}$;
2: **output:** One-slot schedule $\mathcal{S}$;
3: Set c according to Equation (3.18); $\tau := 2 + c$;
4: Sort the links in L in non-decreasing order of length;
5: **for** $v = 1$ **to** n **do**
6:    **if** $a_{l_v}(\mathcal{S}) \leq \tau^{-\alpha}$ **then**
7:       $\mathcal{S} := \mathcal{S} \cup \{l_v\}$;
8:    **end if**
9: **end for**
10: **return** $\mathcal{S}$;

Lemma 3.6 ([42]). Algorithm 2 produces a c-dispersed solution (see Definition 2.10 of a dispered set), where c is defined in Equation (3.18).

Proof. Let l_w be a link in the set $\mathcal{S}$ output by Algorithm 2. Let $\mathcal{S}_w^-$ ($\mathcal{S}_w^+$) be the set of links in $\mathcal{S}$ that are shorter (longer) than l_w. Consider first a link $l_u \in \mathcal{S}_w^-$. Since l_w was added by the algorithm after adding l_u, $a_{l_w}(u) \leq \tau^{-\alpha}$, which implies by Observation 2.5, item 3, that

$$d_{uw} \geq \tau \eta_w^{1/\alpha} d_{ww} > (\tau - 2)\eta_w^{1/\alpha} d_{ww} = c \cdot \eta_w^{1/\alpha} d_{ww}.$$

Consider next a link $l_v \in \mathcal{S}_w^+$. Since l_v was added after l_w, it holds that $a_{l_v}(w) \leq \tau^{-\alpha}$. So by Observation 2.5, $d_{wv} \geq \tau \cdot \eta_v^{1/\alpha} d_{vv}$. Note that $\eta_v \geq \eta_w$ whenever $d_{vv} \geq d_{ww}$. Then, using the triangular inequality,

$$\begin{aligned}
d_{vw} = d(s_v, r_w) &\geq d_{wv} - d_{vv} - d_{ww} \\
&\geq (\tau \eta_v^{1/\alpha} - 2) d_{vv} \\
&\geq (\tau - 2)\eta_w^{1/\alpha} d_{ww} \\
&= c \cdot \eta_w^{1/\alpha} d_{ww}.
\end{aligned}$$

Since this holds for every ordered pair in $\mathcal{S}$, we have that $\mathcal{S}$ is c-dispersed. $\qquad\square$

Lemma 3.7 ([30, 42]). Let l_v be a link in an δ-dispersed set $\mathcal{S}$ (see Definition 2.10 of a dispered set), $\delta \geq 2$. Let $\mathcal{S}_{v\delta}^+$ be the set of links in

$\mathcal{S}$ that are at least as long as l_v. Then,

$$a_{l_v}(\mathcal{S}_{v\delta}^+) < \left(\frac{\alpha - 1}{\alpha - 2}2^5 3\right)\delta^{-\alpha}\eta_v. \tag{3.17}$$

Proof. Our first observation is that disks D_w of radius $\delta \cdot \eta_w^{1/\alpha} \cdot d_{ww} \geq \delta \cdot d_{vv}$ around each receiver $r_w \in \mathcal{S}_{v\delta}^+$ do not contain any sender $s_z \neq s_w$. Using this fact and the triangular inequality, we can lower bound the distance between any two senders $s_w, s_z \in \mathcal{S}_{v\delta}^+$ as $d(s_w, s_z) \geq d(r_w, s_z) - d_{ww} \geq \delta \cdot d_{ww} - d_{ww} = d_{ww}(\delta - 1) \geq d_{vv}(\delta - 1)$. Therefore, disks D_w of radius $d_{vv}(\delta - 1)/2$ around senders in $\mathcal{S}_{v\delta}^+$ do not intersect.

Next, we partition the sender set in $\mathcal{S}_{v\delta}^+$ into concentric rings $Ring_k$ of width $\delta \cdot d_{vv}$ around the receiver r_v. Each ring $Ring_k$ contains all senders $s_w \in \mathcal{S}_{v\delta}^+$, for which $k(\delta \cdot d_{vv}) \leq d_{wv} \leq (k + 1)(\delta \cdot d_{vv})$. We know that the first ring $Ring_0$ does not contain any sender. Consider all senders $s_w \in Ring_k$ for some integer $k > 0$. All disks of radius $d_{vv}(\delta - 1)/2$ around each s_w must be located entirely in an extended ring $Ring_k$ of area:

$$A(Ring_k) = [(d_{vv}(k + 1)\delta + d_{vv}(\delta - 1)/2)^2$$
$$- (d_{vv}k\delta - d_{vv}(\delta - 1)/2)^2]\pi$$
$$= (2k + 1)d_{vv}^2\delta(2\delta - 1)\pi.$$

Since disks D_w of area $A(D_w) \geq (d_{vv}(\delta - 1)/2)^2\pi$ around senders in $\mathcal{S}_{v\delta}^+$ do not intersect, and the minimum distance between r_v and $s_w \in Ring_k, k > 0$ is $k(\delta \cdot d_{vv})$, we can use an area argument to bound the number of senders inside each ring. The total interference coming from ring $Ring_k, k > 1$ is then bounded by

$$I_{l_v}(Ring_k) \leq \sum_{s_w \in Ring_k} I_{l_v}(s_w)$$
$$\leq \frac{A(Ring_k)}{A(D_w)} \cdot \frac{P}{(k\delta d_{vv})^\alpha}$$
$$\leq \frac{(2k + 1)}{k^\alpha} \cdot \frac{4P}{(\delta d_{vv})^\alpha}\frac{\delta(2\delta - 1)}{(\delta - 1)^2}$$
$$\leq \frac{1}{k^{(\alpha - 1)}} \cdot \frac{P}{d_{vv}^\alpha}\frac{2^5 3}{\delta^\alpha},$$

where the last inequality holds since $k \geq 1 \Rightarrow 2k + 1 \leq 3k$ and $\delta \geq 2 \Rightarrow \delta - 1 \geq \delta/2$. Summing up the interferences over all rings yields:

$$I_{l_v}(\mathcal{S}_{v\delta}^{+}) < \sum_{k=1}^{\infty} I_{l_v}(Ring_k)$$

$$\leq \sum_{k=1}^{\infty} \frac{1}{k^{\alpha-1}} \cdot \frac{P}{d_{vv}^{\alpha}} \frac{2^5 3}{\delta^{\alpha}}$$

$$< \frac{\alpha - 1}{\alpha - 2} \cdot \frac{P}{d_{vv}^{\alpha}} \frac{2^5 3}{\delta^{\alpha}}$$

$$= \left(\frac{\alpha - 1}{\alpha - 2} 2^5 3 \right) \delta^{-\alpha} P_{vv},$$

where the last inequality holds since $\alpha > 2$. This results in affectance

$$a_{l_v}(\mathcal{S}_{v\delta}^{+}) = \eta_v \cdot \sum_{l_z \in \mathcal{S}_{v\delta}^{+}} RI_{l_v}(l_z)$$

$$= \eta_v \cdot \frac{I_{l_v}(\mathcal{S}_{v\delta}^{+})}{P_{vv}}$$

$$< \eta_v \cdot \left(\frac{\alpha - 1}{\alpha - 2} 2^5 3 \right) \delta^{-\alpha}. \qquad \square$$

Theorem 3.8 ([42]). Algorithm 2 produces an SINR-feasible solution.

Proof. Let l_w be a link in the set $\mathcal{S}$ output by Algorithm 2. Let $\mathcal{S}_w^{-}$ ($\mathcal{S}_w^{+}$) be the set of links in $\mathcal{S}$ that are shorter (longer) than l_w. The links in $\mathcal{S}_w^{-}$ were processed before l_w, so by the if-condition in the algorithm, $a_{l_w}(\mathcal{S}_w^{-}) \leq \tau^{-\alpha}$. By Lemma 3.6, $\mathcal{S}$ is c-dispersed, so by Lemma 3.7 and the definitions of τ and dispersion,

$$a_{l_w}(\mathcal{S}_w^{+}) < \eta_v \left(\frac{\alpha - 1}{\alpha - 2} 2^5 3 \right) \frac{1}{(c\eta^{1/\alpha})^{\alpha}}$$

$$= \frac{1}{c^{\alpha}} \left(\frac{\alpha - 1}{\alpha - 2} 2^5 3 \right)$$

$$\leq \frac{1}{2}, \quad \text{where}$$

$$c = \max\left(2, \left(\frac{\alpha - 1}{\alpha - 2} 2^6 3\right)^{\frac{1}{\alpha}}\right). \tag{3.18}$$

We have shown that the affectance of each link in $\mathcal{S}$ is at most $(\tau^{-\alpha} + 1/2) < 1$, which means that $SINR_{l_v} \geq \beta$ for every scheduled link $l_v \in \mathcal{S}$. This concludes the proof of the lemma. $\qquad\square$

3.3.1.2 Approximation Ratio

We start the performance analysis of Algorithm 2 with two definitions and a combinatorial lemma, to which we refer as the *blue-dominant centers lemma*. Informally, if we are given two sets of points, let's call them red and blue points, we say that a blue point is *blue-dominant* if it is "shadowed", or "protected", by other blue points from the red points in all directions. We call the set of blue points that "protect" the blue-dominant point from the red points a *guarding set*.

Definition 3.9. Let $\mathcal{R}$ and $\mathcal{B}$ be two disjoint sets of points in a metric space $(\mathcal{V}, d)$. Let's call them *red* and *blue* points, respectively. For q a positive integer, a point $b \in \mathcal{B}$ is *q-blue-dominant* if every ball $B_\delta(b)$ around b, comprising points w such that $d(w, b) \leq \delta$, contains $q \in \mathbb{Z}^+$ times more blue points than red points. Formally,

$$\forall \delta \in \mathrm{R}_0^+ : |B_\delta(b) \cap \mathcal{B}| > q \cdot |B_\delta(b) \cap \mathcal{R}|.$$

Definition 3.10. Let $\mathcal{R}$ and $\mathcal{B}$ be defined as above. Let $r \in \mathcal{R}$ be a red point and $G(r) \subseteq \mathcal{B}$ be a set of blue points. We say that $G(r)$ is *guarding* r if for all $b^* \in \mathcal{B} \setminus G(r)$, we have that $B_{d(b^*, r)}(b^*) \cap G(r) \neq \emptyset$. Furthermore, we say that $G_q(r)$ is *q-guarding* r if for all $b^* \in \mathcal{B} \setminus G_q(r)$, we have that $B_{d(b^*, r)}(b^*) \cap G_q(r) \geq q$.

Lemma 3.11. ([30] *Blue-dominant centers lemma*) Let $\mathcal{R}$ and $\mathcal{B}$ be two disjoint sets of red and blue points in the two-dimensional

Euclidean space, and q be a positive integer. If $|\mathcal{B}| > 6q|\mathcal{R}|$ then there exists at least one q-blue-dominant point in $\mathcal{B}$.

Proof. Process the points in $\mathcal{R}$ in an arbitrary order while maintaining a subset $\mathcal{B}'$ of $\mathcal{B}$ as follows (initially, $\mathcal{B}' = \mathcal{B}$). For $r \in \mathcal{R}$ construct a q-guarding set $G_q(r) \subseteq \mathcal{B}'$ (guarding r relative to the current $\mathcal{B}'$) and remove $G_q(r)$ from $\mathcal{B}'$.

We claim that it is possible to construct a guarding set $G_q(r)$ of size at most $6q$. The procedure to construct $G_q(r)$ is as follows (see Figure 3.4). Consider a red point r. Draw six sectors of $60°$ originating at r. For each of these six sectors sec_j, include the closest q blue points $b_j \in sec_j$ in $G_q(r)$ (if sec_j has less than q blue points from $\mathcal{B}'$, pick all the blue points in this sector). Now $G_q(r)$ has size at most $6q$, and we claim that it is guarding r. Suppose it is not. Then there is a point $b^* \in \mathcal{B}' \setminus G_q(r)$ with $B_{d(b^*,r)}(r) \cap G_q(r) < q$. Suppose b^* is located in sec_j and we selected q blue points b_j from sec_j into $G_q(r)$. This means that $d(b^*, b_j) > d(b^*, r)$ for some $b_j \in sec_j$, which implies that the sector angle is larger than $60°$. (Note that if $G_q(r)$ contains less than q blue points b_j from sector sec_j, then b^* would have been picked to guard r in that sector, also establishing a contradiction.)

After going through all points in $\mathcal{R}$, the set $\mathcal{B}'$ is still nonempty by the assumption on the relative sizes of $\mathcal{R}$ and $\mathcal{B}$. We claim that every

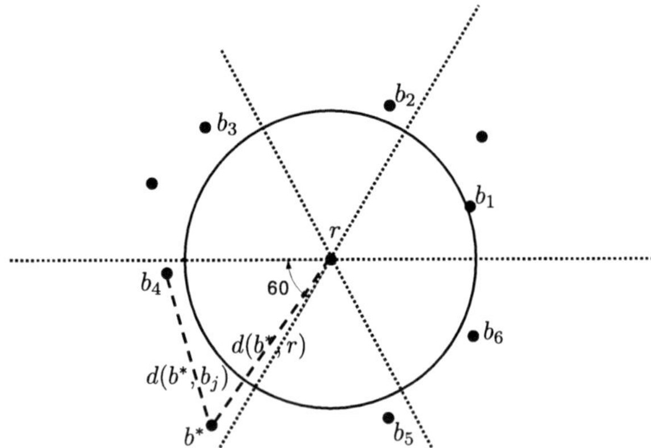

Fig. 3.4 Constructing a q-guarding set $G_q(r), q = 1$ of size at most $6 \cdot q = 6$ for the red point r ($G_q(r) = \{b_1,\ldots,b_6\}$).

point $b^* \in \mathcal{B}'$ is now q-blue-dominant. This holds since (1) all $G_q(r)$'s are pairwise disjoint and (2) every ball $B_\delta(b^*), b^* \in \mathcal{B}'$, that contains a red point r, contains also q blue points. Hence, for every blue node $b^* \in \mathcal{B}'$, every ball $B_\delta(b^*)$ contains q times more blue points than red points ("more", since the center b^* is also blue). $\qquad\square$

In the following lemma the algorithm's solution is compared with the stricter optimal solution OPT_p. The lemma applies dispersion robustness properties of the physical model, discussed in Section 2.4.2, and the result of Lemma 3.11.

Lemma 3.12 ([42]). Let OPT_p be a p-signal (one-slot) optimal schedule (see Definition 2.6 for OPT_p), where $p = \tau^\alpha$, $\tau = 2 + c$ and c is defined in Equation (3.18). Let $\mathcal{S}$ be the set of links scheduled by Algorithm 2. Then, $|OPT_p| \le \rho|\mathcal{S}|$, where $\rho = 6(q/\tau)^\alpha$ and $q = 2 + \tau$.

Proof. Let $OPT_p' = OPT_p \setminus \mathcal{S}$ and $\mathcal{S}' = \mathcal{S} \setminus OPT_p$. Let $\mathcal{B} = \{s_v | \ell_v \in OPT_p'\}$ and $\mathcal{R} = \{s_w | \ell_w \in \mathcal{S}'\}$ be the sets of senders in OPT_p' and $\mathcal{S}'$; we call them blue and red points, respectively. Suppose the claim is false. By Lemma 3.11, there is a $(q/\tau)^\alpha$-blue-dominant point (sender) s_b^* in $\mathcal{B}$. We shall argue that the link $l_b^* = (s_b^*, r_b^*)$ would have been picked up by Algorithm 2, which is a contradiction.

Consider any red point $s_r \in \mathcal{R}$. Let $G^*(s_r)$ be the set of points (senders) in s_r's $(q/\tau)^\alpha$-guarding set that are closer to s_b^* than s_b^* is to s_r. They are all within radius $d(s_b^*, s_r)$ from s_b^*. By the blue-dominant center property, $|G^*(s_r)| \ge (q/\tau)^\alpha$. By Lemma 2.14, since OPT_p' is a p-signal set and $p = \tau^\alpha$, OPT_p' is then τ-dispersed. Using this fact and Lemma 2.11, we know that $d(s_r, r_b^*) \ge q \cdot (\eta_b^*)^{1/\alpha} \cdot d(s_b^*, r_b^*) \ge q \cdot d(s_b^*, r_b^*)$. By the triangular inequality, it then follows that

$$d(s_b^*, r_b^*) \le \frac{d(s_r, r_b^*)}{q}$$

$$\le \frac{d(s_r, s_b^*) + d(s_b^*, r_b^*)}{q} \quad \Rightarrow$$

$$d(s_b^*, r_b^*) - \frac{d(s_b^*, r_b^*)}{q} \leq \frac{d(s_b^*, s_r)}{q} \quad \Rightarrow$$

$$d(s_b^*, r_b^*) \leq \frac{d(s_b^*, s_r)}{(q-1)}.$$

Which results in

$$d(s_r, r_b^*) \geq d(s_b^*, s_r) - d(s_b^*, r_b^*)$$

$$\geq d(s_b^*, s_r) \cdot \frac{(q-2)}{(q-1)}.$$

Moreover, for each $s_b \in G^*(s_r)$,

$$d(s_b, r_b^*) \leq d(s_b^*, s_b) + d(s_b^*, r_b^*)$$

$$\leq d(s_b^*, s_r) + d(s_b^*, r_b^*)$$

$$\leq d(s_b^*, s_r) \cdot \frac{q}{(q-1)}.$$

The affectance of the red sender s_r on blue receiver r_b^* is then bounded by

$$a_{r_b^*}(s_r) = \eta_b^* \cdot \frac{d(s_b^*, r_b^*)^\alpha}{d(s_r, r_b^*)^\alpha}$$

$$\leq \eta_b^* \cdot \left(\frac{q-1}{q-2}\right)^\alpha \cdot \frac{d(s_b^*, r_b^*)^\alpha}{d(s_b^*, s_r)^\alpha}.$$

In comparison, the combined affectance of the blue senders $s_b \in G^*(s_r)$ on r_b^* is at least

$$a_{r_b^*}(G^*(s_r)) = \eta_b^* \cdot \sum_{s_b \in G^*(s_r)} \frac{d(s_b^*, r_b^*)^\alpha}{d(s_b, r_b^*)^\alpha}$$

$$\geq \eta_b^* \cdot \left(\frac{q}{\tau}\right)^\alpha \cdot \left(\frac{q-1}{q}\right)^\alpha \cdot \frac{d(s_b^*, r_b^*)^\alpha}{d(s_b^*, s_r)^\alpha}$$

$$\geq \left(\frac{q-2}{\tau}\right)^\alpha \cdot a_{r_b^*}(s_r)$$

$$= a_{r_b^*}(s_r).$$

Since this holds for any $s_r \in \mathcal{R}$, the total interference that r_b^* receives from blue senders (those in OPT_p') is at least as high as the interference it would receive from the red senders (those in $\mathcal{S}'$):

$$a_{r_b^*}(\mathcal{S}) = a_{r_b^*}(\mathcal{S}') + a_{r_b^*}(\mathcal{S} \cap OPT_p)$$

$$< a_{r_b^*}(OPT_p') + a_{r_b^*}(\mathcal{S} \cap OPT_p)$$

$$= a_{r_b^*}(OPT_p)$$

$$\leq \tau^{-\alpha}.$$

Since link l_b^* is in OPT_p, receiver r_b^* is affected by at most $1/p = \tau^{-\alpha}$ by senders in OPT_p. Since the affectedness of r_b^* by the red senders is less than the affectance caused by blue senders, and therefore less than $\tau^{-\alpha}$, link l_b^* would have been picked up by Algorithm 2 for the slot i_0, which establishes the contradiction. $\qquad\square$

The following result follows from Lemma 3.12 in combination with the correctness result in Theorem 3.8 and the robustness Corollary 2.9.

Theorem 3.13. Algorithm 2 approximates the One-Slot Scheduling Problem within a constant factor.

3.3.2 $O(\log n)$-Approximation Algorithm for the Multi-slot Scheduling Problem

In this section we show that, if Algorithm 2 is applied repeatedly, it provides an $O(\log n)$ approximation for the Multi-slot Scheduling Problem.

Theorem 3.14. If Algorithm 2 is applied repeatedly to schedule all links, the resulting schedule is an $O(\log n)$ approximation for the Multi-slot Scheduling Problem.

Proof. Suppose the instance is schedulable in t slots, and we find in each iteration an ρ-approximation to the One-Slot Scheduling Problem. Then, this ρ-approximate set is of size at least $n/(t\rho)$, where n is the current number of links. In other words, the size of the remaining instance is at most $n(1 - 1/(t\rho))$. After s iterations, the size is at most $n(1 - 1/(t\rho))^s \leq ne^{-s/(t\rho)}$. Plugging in $s = t\rho\log n$, we are down to a single link. This gives an approximation ratio of $s/t = \rho \log n = O(\log n)$. $\qquad\square$

3.3.3   Handling Different Transmission Powers

We can treat the case when links transmit with different powers in two different ways. Let $P_{\max}$ ($P_{\min}$) be the maximum (minimum) power used by a link, respectively. By introducing a factor of $P_{\min}/P_{\max}$ into the affectance threshold $\tau^{-\alpha}$, Algorithm 2 still produces a feasible schedule, that is longer by a factor of at most $P_{\max}/P_{\min}$.

Alternatively, we can partition the instance into "power regimes", where each regime consists of links whose powers are equal up to a factor of 2. We schedule each power regime separately, obtaining an approximation factor of at most $\log P_{\max}/P_{\min}$, or at most the number of different power values.

Kesselheim and Vöcking [54] propose a distributed algorithm for multi-slot scheduling with fixed power assignments. The algorithm achieves a $O(\log^2 n)$-approximation ratio. Each sender transmits its message with a certain probability and stops as soon as it obtained an acknowledgment message. The power assignments have to satisfy a monotonicity requirement and the transmission power for sending acknowledgments needs to be adjustable. For the special case of the uniform power assignment the last requirement can be dropped.

3.4   Outlook

In this chapter we studied the problem of scheduling wireless links when the uniform power assignment is used, i.e., all nodes transmit with the same power level. As pointed out in Section 3.3.3, the asymptotic bounds on the performance of the algorithms presented in Sections 3.2 and 3.3 hold also in a scenario where nodes have different, but fixed power levels, as long as the ratio $P_{\max}/P_{\min}$ is bounded by a constant, or if there is a constant number of power levels.

We showed in Section 3.1 that the Multi-slot Scheduling Problem is NP-hard in the physical interference model, and presented a simple algorithm to solve the problem with a guaranteed approximation ratio in Section 3.2. The optimality guarantee of this first algorithm, however, is dependent on the topology of the network, and unfortunately can get as bad as $\Omega(n)$ in some special cases. In Section 3.3 we presented considerably improved an algorithm, whose approximation ratio no longer depends on the topology of the network.

Some problems, however, still remain open in this context, e.g., whether there exists a polynomial-time approximation scheme (PTAS) for any of the scheduling problems, i.e., an algorithm that takes an instance of the problem and a parameter $\varepsilon > 0$ and, in polynomial time, produces a solution that is within a factor ε of the optimum.

In the next chapter, we will address the scheduling problems *with* power control, i.e., a scenario where nodes can adjust their power levels to achieve better throughput.

4

Scheduling With Power Control

So far we have considered network instances where all devices emit signals of the same power level. However, there exist devices with hardware that can adjust the transmission power. While this increases the complexity of the devices, the throughput can be higher if the power levels of simultaneously transmitting devices differ. In this chapter, we examine the benefit of power control as well as its limitations.

We examine oblivious power assignments, where the power only depends on the distance a signal has to cover, in Section 4.3. While these assignments are useful for the design of distributed algorithms, they exhibit the shortcoming that they can lead to a performance loss in the order of the size of the network. In Section 4.4 we study more general power assignment strategies and give bounds on their performance.

4.1 The Power of Power Control

We start with illustrative examples from Ref. [67]. In the first example two sender receiver pairs (s_1, r_1) and (s_2, r_2) are arranged on a line such that both pairs transmit in the same direction and their transmission lines do not cross. This setup is shown in Figure 4.1. The question

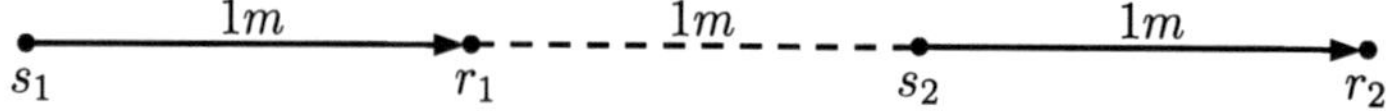

Fig. 4.1 Example with four nodes on a line with ambient noise of 0.01 μW and a reception threshold $\beta = 3$. When s_1 sends with $P(s_1) = 1$ dBm and s_2 with $P(s_2) = -8$ dBm, we obtain signal-to-noise and interference ratios of $\frac{1.26 \text{ mW}/1^3}{0.01 \text{ }\mu\text{W}+(0.16 \text{ nW}/1^3)} \approx 8.12$ at r_1 and $\frac{0.16 \text{ mW}/1^3}{0.01 \text{ }\mu\text{W}+(1.26 \text{ mW}/3^3)} \approx 3.32$ at r_2. That is, the SINR threshold is exceeded for both pairs, thus node r_1 can perfectly decode s_1's message, and at the same time, r_2 successfully receives s_2's message. There is no collision.

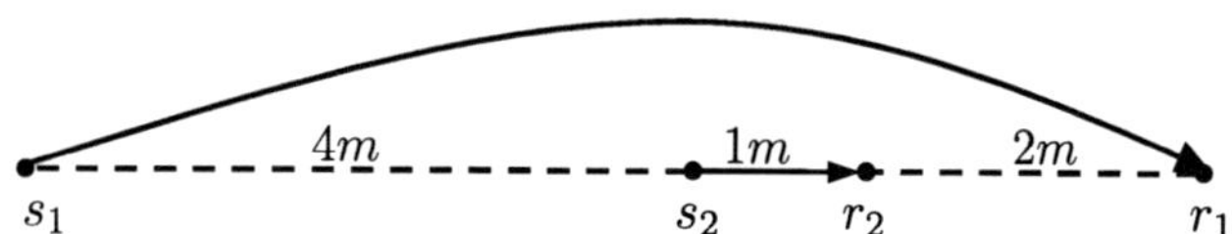

Fig. 4.2 Example with two nested links, where one link's sender and receiver are in the transmission line of another communication link. When s_1 sends with $P(s_1) = 1$ dBm and s_2 with $P(s_2) = -15$ dBm, we get an SINR of $\frac{1.26 \text{ mW}/7^3}{0.01 \text{ }\mu\text{W}+(31.6 \text{ }\mu\text{W}/3^3)} \approx 3.11$ at r_1 and $\frac{31.6 \text{ }\mu\text{W}/1^3}{0.01 \text{ }\mu\text{W}+(1.26 \text{ mW}/5^3)} \approx 3.13$ at r_2. That is, the SINR threshold is exceeded for both pairs, thus node r_1 can perfectly decode s_1's message, and at the same time, r_2 successfully receives s_2's message. There is no collision.

is whether it is necessary to schedule the two messages in succession or if they can be sent in the same time slot without colliding at any of the two receivers. Clearly, if both senders emit signals of the same power, trying to send the two messages in parallel will fail because both senders s_1 and s_2 are at the same distance from r_1. If the powers can be adjusted, however, both messages can easily be transmitted simultaneously, thereby doubling the achieved throughput.

By rearranging the two sender–receiver pairs (s_1, r_1) and (s_2, r_2) we obtain a setting where one pair is placed in the transmission line of the other (nested links). This setup is shown in Figure 4.2. As before, the question is whether it is really necessary to schedule the two messages in succession or if they can be sent in the same time slot without colliding at any of the two receivers. Clearly, any graph-based approach trying to send the two messages in parallel will fail because, intuitively, the medium between s_2 and r_2 can only be used once per time slot.

In the physical model, however, both messages can easily be transmitted simultaneously, thereby doubling the achieved throughput.

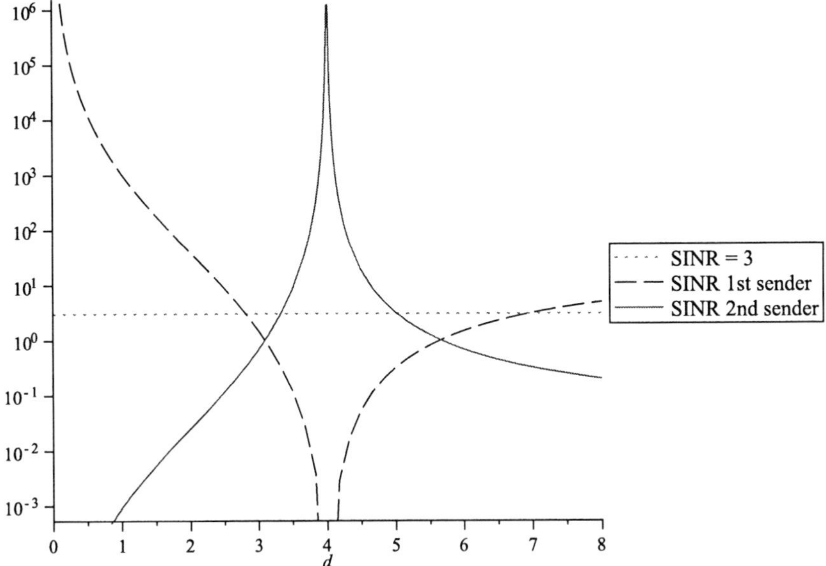

Fig. 4.3 This figure shows the change of the SINR depending on the receiver location. The sender s_1 located at position 0 and sender s_2 at position 4. s_1 sends with an output power of 1 dBm while s_2 sends with -15 dBm. The path-loss exponent α is equal to 3. The dotted line represents the reception level β. This illustrates that the signal of the first sender can be decoded between 0 and 2.8 as well as at 6.9 and further away from 0. The signal of the second sender can be received successfully between 3.3 and 4.9.

To visualize this result, Figure 4.3 depicts how the SINR along the line on which the four nodes are placed changes. It is clearly visible that P_{s_2} dominates P_{s_1} in the proximity of r_2, but diminishes faster than P_{s_1} when approaching r_2. Note that for a successful parallel transmission, it is crucial that the two senders do not utilize the same output power.

The example of two nested links can be extended by adding more links enclosing the two present links illustrated in Example 4.4 and Figure 4.4. If they are not too close together, a power assignment exists that allows the simultaneous transmission of all links. With uniform power, every link has to be scheduled separately. In other words, power control can increase the capacity from one to n, depending on the locations of the links.

The first work quantifying how much power control can improve the capacity of wireless networks describes is due to Moscibroda et al. [66].

The authors show that uniform and linear (see Section 4.3.2) power assignments can be at least $\Omega(n/\log^4 n)$ times worse than optimal. Before examining various classes of power assignments, we discuss in the next section how we can find the best power assignment for a given set of communication requests.

4.2 Feasibility

Remember that in the physical model the propagation attenuation (or link gain) between a sender node s_i and a receiver node r_j in the Euclidean plane is modeled as $g(s_i, r_j) = d(s_i, r_j)^{-\alpha}$. Whether or not a set of links can be scheduled in the same time slot depends on the link gain between all sender–receiver pairs. Note that the link gain matrix

$$Z = \left[\frac{g(s_j, r_i)}{g(s_i, r_i)} \right]_{i,j} = \left[\frac{d(s_i, r_i)^\alpha}{d(s_j, r_i)^\alpha} \right]_{i,j}$$

is a matrix consisting of positive values only. Zander [86] showed that this property can be exploited to compute the *maximum achievable* SIR^* for wireless networks efficiently.[1] Finding a power assignment yielding the maximal SIR level can essentially be reduced to solving an Eigenvalue problem for the link gain matrix Z. Due to results by Perron, Frobenius, and Wielandt [25] on the theory of non-negative matrices we can deduce that for positive matrices:

(1) there is exactly one real eigenvalue λ^* for which all elements of the corresponding eigenvector have the same sign; and

(2) the minimum real λ such that the inequality

$$\lambda P \geq ZP$$

holds for $P \geq 0$ is $\lambda = \lambda^*$.

This implies that the maximum achievable SIR^* is given by

$$SIR^* = \frac{1}{\lambda^* - 1}.$$

---

[1] Zander does not take noise into account, thus his results are valid in the *Signal-to-Interference-Ratio* (*SIR*) model mentioned in Section 2. See Refs. [10, 71] for the treatment of ambient noise level and varying SINR ratio requirements among the nodes.

Furthermore, the corresponding eigenvector $\mathbf{P}^*$ constitutes a power vector reaching this maximum for all links, i.e., they all have the same *SIR* level.

Theorem 4.1 ([86]). In the absence of noise, a set of links can be successfully scheduled in one time slot if and only if the largest eigenvalue of the link gain matrix is less than $1/\beta + 1$.

This result can be used to determine whether a set of links can be scheduled in one time slot, i.e., whether the set is *feasible* and what the highest achievable *SIR* level is. Since solving a Eigenproblem takes time in $O(n^3)$, this time complexity carries over for Theorem 4.1.

Inevitably there is noise in every real system. Nonetheless, this result is still useful since it provides a efficient method to determine if a set of links can NOT be scheduled concurrently. The converse however does not hold in the physical model. However, it is possible to extend this approach to take noise into account. Borbash and Ephremides [10] and Pillai et al. [71] demonstrate how the best possible power assignment can be computed when the ambient noise level and the required SINR ratio vary among the nodes.

4.2.1 Note on Bounded Resources

The power assignment computed by solving this eigenvalue problem might require that transmission power levels differ by a factor exponential in n. In reality however, every device has limited capabilities to adjust its transmission power, typically between -10 dBm and 10 dBm [79]. Furthermore, if we look at some of the smallest and largest energy generators known to mankind, the power a cell of the human body produces/consumes is in the order of 1 pW, whereas the hydroelectric power plant at the Assuan dam produces around 127 MW, thus the number of magnitudes of possible power levels for wireless devices is clearly a small constant far below 100. The area where the nodes are distributed is typically constrained as well. Avin et al. quantify in Ref. [5] the worst-case performance loss caused by applying a uniform power assignment, compared to an optimal power assignment. They show that the performance gain of power control is at most in $O(\log \frac{P_{\max}}{P_{\min}})$ and $O(\log \frac{d_{\max}}{d_{\min}})$.

4.3 Oblivious Power Assignment

The previous sections have demonstrated that: (a) power control can increase the capacity of a network enormously, even by a factor of n in worst-case scenarios and (b) we can compute the best possible signal-to-noise and interference ratio and the corresponding power assignment efficiently. However, in a typical network the nodes do not know how many other nodes intend to transmit at the same time nor their locations. They have to choose a signal strength independently. Clearly, the uniform power assignment is the easiest choice, however it cannot tap the full potential of the wireless channel. In order to still exploit the capacity gain power control offers compared to uniform power, power assignments that are only based on the length of the links have been proposed. Such *oblivious power assignments* [22] do not need information on the location of senders and receivers nor on potential concurrent transmissions. Oblivious power assignments lead to simple distributed algorithms that decide whether to send or not in a randomized manner. However, the drawback of these algorithms is the fact that every one of them can lead to an arbitrarily bad performance, as shown in Section 4.3.1.

On the positive side, Halldórsson [40] showed that any constant approximation scheduling algorithm for the Multi-slot Scheduling Problem adopting an oblivious power assignment gives an $O(g(L))$ approximation of Multi-slot Scheduling with power control. $O(g(L))$ is the link diversity, the number of different length classes of L, as defined in Definition 2.1. This result is due to the fact that it is not possible to achieve a schedule length more than a constant factor shorter when scheduling links of lengths differing by a factor of less than 2 with power control. In other words, if the links are roughly of the same length, one may as well use uniform transmission power and apply any scheduling algorithm with a constant approximation ratio. E.g., one could use Algorithm 2.

Theorem 4.2 ([40]). If the link lengths differ by at most a factor of two, then there exists an $O(1)$-approximation algorithm for the Multi-slot Scheduling Problem with power control using any oblivious transmission power assignment.

To handle links of variable lengths, we can group them into $O(g(L))$ classes, based on their length and schedule each class separately (Theorems 3.4 and 3.5 in Ref. [40]).

Theorem 4.3 ([40]). The Multi-slot Scheduling Problem with power control is $O(g(L))$-approximable with any oblivious transmission power assignment.

Thanks to a reduction of a scheduling instance to a Unit Disk Graph (UDG) in Ref. [40], algorithms for coloring of UDGs can be applied and their complexities carry over. As a consequence, the online scheduling problem as well as the distributed scheduling problem with power control are $O(g(L))$-approximable.

A widely used example [6, 83] of an oblivious power assignment strategy is the *energy metric* or *linear power assignment*, where the transmission power of a link with length d is proportional to the energy loss, i.e., in the order of d^α. Another assignment strategy that caught much attention recently is the *square root* or *mean power assignment*, where the transmission power is set to $\sqrt{d}$. These assignment strategies are studied in detail in Sections 4.3.2 and 4.3.3. With the square root assignment, one can prove that there is an algorithm that constructs schedules that are a factor of $O(\log \log \Lambda \log n)$ longer than an optimal algorithm with an arbitrary power assignment [40]. Depending on the given set of links L, this can be faster or much slower than the bound $O(g(L))$.

4.3.1   Oblivious Lower Bound

Fanghänel et al. [22] were the first to use the term *oblivious power assignment* to include all power assignment schemes that set the transmission power for a link based on a function of the distance between the nodes of the link. More precisely, an oblivious power assignment requires a function $f\colon \mathbb{R}_{>0} \to \mathbb{R}_{>0}$ such that for every link $l_i = (s_i, r_i)$, $P(s_i) = f(d(s_i, r_i))$.

Unfortunately, the resulting schedules can be arbitrarily bad compared to the optimal schedule with an arbitrary power assignment. In fact, Moscibroda et al. [66] show that for uniform and linear power assignments there exists an instance with n communication requests requiring $\Omega(n)$ time slots even though a different power assignment schedules the requests in $O(\log^4 n)$ time slots. This construction has been extended by Fanghänel et al. [22] for a $\Omega(n)$ lower bound for all oblivious power assignments. In other words, oblivious assignments cannot yield approximation ratios better than $\Omega(n)$ for the scheduling problem, which corresponds to the worst possible performance guarantee.

Before we study this problem for arbitrary oblivious power assignment functions, let us briefly examine a one-dimensional problem instance from [65]. For this instance, we can easily verify that the most popular oblivious power assignment approaches, uniform and linear assignments do lead to bad schedules.

The first problem instance consists of n nested links of exponentially growing length as illustrated in Figure 4.4.

Example 4.4 (Exponential Nesting). All the sender and receiver nodes are situated on a straight line with the following distance to 0: sender node $s_i = -x^{i-1}$, receiver node $r_i = x^{i-1}, \forall 0 < i \leq n$. Hence the length of the ith link is $d(s_i, r_i) = 2x^{i-1}$. See Figure 4.4.

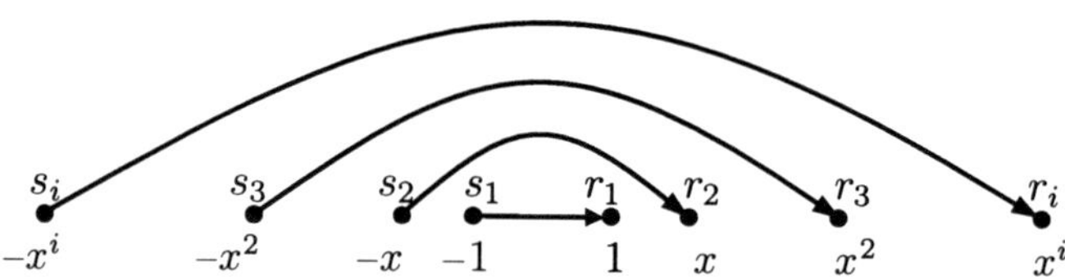

Fig. 4.4 A set of communication requests of exponentially increasing length. For this instance, uniform and linear power assignments yield a schedule of length n whereas one time slots suffice under the square root assignment.

Proposition 4.1. Uniform and linear power assignments schedule nested links separately, even though they can be transmitted simultaneously if the constant x is set to $2^{2+2/\alpha}\beta^{2/\alpha}$.

Proof. In a feasible set of links under a uniform power assignment, the ball with radius $d(s_i, r_i)$ around receiver node r_i cannot contain any other sender s_j, $j \neq i$. In the nested scenario the longer link of any pair of links suffers from interference that is greater than the received power and can thus not decode its message correctly. This is the reason why any algorithm scheduling the nested link set with uniform power needs n slots.

Let us now determine the interference caused by sender s_j at the receiver r_i, where $i < j$ under the linear power assignment.

$$I_{r_i}(s_j) = \frac{P(s_i)}{d(s_i, r_1)^\alpha} = \left(\frac{2x^{j-1}}{x^{i-1} + x^{j-1}} \right)^\alpha > 1.$$

Since the received power of sender s_i at r_i is exactly 1, the ratio $\frac{P_{r_i}(s_i)}{I_{r_i}(s_j)}$ is less than 1 for any link l_j, where $j > i$. As a consequence, we have to schedule every link l_i in a time slot without any concurrent transmissions.

It remains to prove that a constant number of time slots suffice to schedule n nested links. To this end, we apply the square root assignment, where link l_i is assigned a power level of $P(s_i) := \sqrt{d(s_i, r_i)^\alpha} = (2x^{i-1})^{\alpha/2}$. Hence the received power at r_i is:

$$P_{r_i}(s_i) = \frac{(2x^{i-1})^{\alpha/2}}{(2x^{i-1})^\alpha} = (2x^{i-1})^{-\alpha/2}.$$

The interference caused by sender s_j at receiver r_i amounts to

$$I_{r_i}(s_j) = \frac{P(s_j)}{d(s_j, r_i)^\alpha} = \frac{(2x^{j-1})^{\alpha/2}}{(x^{j-1} + x^{i-1})^\alpha}.$$

For a given link l_i, the set of shorter links $L_j^- = \{l_j | j < i\}$ is responsible for interference

$$I_{r_i}(L_j^-) = \sum_{j=1}^{i-1} I_{r_i}(s_{i-j})$$

$$= \sum_{j=1}^{i-1} \frac{(2x^{j-1})^{\alpha/2}}{(x^{j-1} + x^{i-1})^\alpha}$$

$$< 2^{\alpha/2} \sum_{j=1}^{i-1} x^{-\alpha/2(i-j-1)}$$

$$= 2^{\alpha/2} x^{-(i-1)\alpha/2} \sum_{j=1}^{\infty} x^{-j\alpha/2}$$

$$\leq 2^{\alpha/2+1} x^{-(i-1)\alpha/2-\alpha/2}.$$

The last inequality holds since $\sum_{j=1}^{\infty} c^j = c/(1 - c) < 2c \; \forall |c| < 1/2$. For the set of longer links $L_j^+ = \{l_j | j > i\}$, similar arguments imply that

$$I_{r_i}(L_j^+) = \sum_{j=1}^{n} I_{r_i}(s_{i+j})$$

$$< \sum_{j=1}^{\infty} 2^{\alpha/2} x^{-(i+j-1)\alpha/2}$$

$$\leq 2^{\alpha/2+1} x^{-(i-1)\alpha/2-\alpha/2}.$$

Hence, the SINR condition is satisfied for all links l_i:

$$SINR_{r_i}(L) = \frac{P_{r_i}(s_i)}{I_{r_i}(L_j^-) + I_{r_i}(L_j^+)} > \frac{(2x^{i-1})^{-\alpha/2}}{2^{\alpha/2+2} x^{-(i-1)\alpha/2-\alpha/2}}$$

$$= \frac{x^{\alpha/2}}{2^{\alpha/2+2}} \geq \beta. \qquad \square$$

This example is useful to quickly test, whether a given algorithm is able to schedule "difficult" instances well. A more general family of instances shows that all oblivious power assignment functions lead to long schedules.

Theorem 4.5 ([22]). Let $f : \mathbb{R}_{>0} \to \mathbb{R}_{>0}$ be any oblivious power assignment function. There exists a family of instances on a line that requires $\Omega(n)$ time slots for the Multi-slot Scheduling Problem when transmitting with power levels defined by f, whereas an optimal power control schedule has constant length.

Proof. We prove that given an oblivious power assignment function f, we can place n sender–receiver pairs in a way that only a constant number of pairs can be scheduled concurrently when using this power assignment. With a different power assignment, all senders can transmit simultaneously.

We distinguish three cases, depending on the asymptotic behavior of f.

(a) f is asymptotically unbounded: for every $c > 0$ and every $x_0 > 0$ there exists a value $x > x_0$ with $f(x) > c$.

(b) f is asymptotically bounded from above, but does not converge to 0: There is a value $b \in (0, c]$ such that for all $x_0 > 0$ there exists a value $x > x_0$ with $b/2 \leq f(x) \leq b$.

(c) f converges to 0: $\lim_{x \to \infty} = 0$.

For the cases (b) and (c), we can use Example 4.4 and generalize Proposition 4.1. This is possible due to the fact that f is almost uniform in these cases. Hence a schedule assigning a power level of $f(2x^{i-1})$ to the sender s_i uses $\Omega(n)$ time slots, even though all links can be transmitted concurrently.

For case (a), the example from Ref. [66] is generalized as follows. The nodes are positioned on the line, such that all pairs transmit to the right and no nestings or crossings occur, an example is depicted in Figure 4.5. We define an arrangement of the nodes by the length

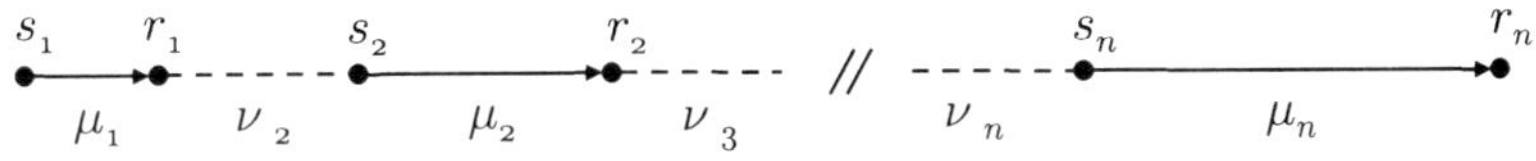

Fig. 4.5 Lower bound instance for oblivious assignment. The distances between the nodes depend on f.

$\mu_i = r_i - s_i$ of link l_i and the distance ν_i between receiver node r_{i-1} and sender node s_i. We set the first sender node at position 0.

Formally, this kind of instance can be defined by $s_1, r_1, s_2, r_2, \ldots, s_n, r_n \in \mathbb{R}$ and $\nu_1, \mu_1, \nu_2, \mu_2, \ldots, \nu_n, \mu_n$ such that

$$s_i = \begin{cases} 0 & \text{if } i = 1 \\ r_{i-1} + \nu_i & \text{otherwise} \end{cases},$$

and

$$r_i = \begin{cases} 1 & \text{if } i = 1 \\ r_i = s_i + \mu_i & \text{otherwise} \end{cases}.$$

We set $\nu_1 := 0$, $\mu_1 := 1$, $\tau := (2\beta + 1)^{2/\alpha}$ and we define the distances ν_i and μ_i between the nodes recursively: Given $\mu_1, \ldots, \mu_{i-1}$ and ν_i, we choose ν_i and μ_i such that

$$\nu_i := \tau(\nu_{i-1} + \mu_{i-1}), \tag{4.1}$$

$$\mu_i \geq \nu_i \quad \wedge \quad f(\mu_i) \geq \nu_i^\alpha \max_{j<i}\left(\frac{f(\mu_j)}{\mu_j^\alpha}\right). \tag{4.2}$$

Since f is asymptotically unbounded, it is always possible to find μ_i satisfying Equation (4.2). As a consequence of Equations (4.1) and (4.2) it holds that $\nu_i > \tau\mu_{i-1} \geq \tau\nu_{i-1}$ and hence

$$\nu_i > \tau^{i-j}\nu_j \; \forall \; 0 \leq j \leq i. \tag{4.3}$$

Thanks to this construction ν_i grows exponentially. This fact guarantees that the receiver of a link l_k is exposed to high interference by links with higher indices. Let $L' \subset L$ be a set of links that can be transmitted simultaneously and let k be the lowest index among the links in L'. Due to Equation (4.1) it holds for $i \in L' \backslash l_k$ that

$$d(s_i, r_k) = \sum_{j=k+1}^{i-1} \mu_j + \sum_{j=k+1}^{i} \nu_j$$

$$\overset{(4.1)}{\leq} 2\sum_{j=k}^{i} \nu_j$$

$$\stackrel{(4.3)}{\leq} 2\sum_{j=k}^{i} \tau^{-(i-j)}\nu_i$$

$$\leq 2\nu_i \sum_{j=0}^{\infty} \tau^{-j}$$

$$= \frac{2\nu_i}{1-\tau}. \tag{4.4}$$

The last transformation holds since $\sum_{j=0}^{\infty} c^j = 1/(1-c)$. Combining Equations (4.2) and (4.4) we obtain the following bound for interference caused at r_k by the set L'.

$$I_{r_k}(L') = \sum_{i\in L'\setminus\{l_k\}} \frac{P(s_i)}{d(s_i, r_k)^\alpha}$$

$$\stackrel{(4.4)}{\geq} \sum_{i\in L'\setminus\{l_k\}} \frac{f(\mu_i)}{\left(\frac{2\nu_i}{1-\tau}\right)^\alpha}$$

$$\stackrel{(4.2)}{\geq} \sum_{i\in L'\setminus\{l_k\}} \frac{\nu_i^\alpha f(\mu_k)(1-\tau)^\alpha}{\mu_k^\alpha (2\nu_i)^\alpha}$$

$$= \frac{(|L'|-1)f(\mu_k)(1-\tau)^\alpha}{\mu_k^\alpha 2^\alpha}. \tag{4.5}$$

Since the set L' is feasible by definition, the SINR condition for l_k is satisfied

$$SINR_{r_k}(L') = \frac{P_{r_k}(s_k)}{I_{r_k}(L')} \stackrel{(4.5)}{>} \frac{2^\alpha}{(|L'|-1)(1-\tau)^\alpha} \geq \beta.$$

As a consequence $|L'| \leq \frac{2^\alpha}{(1-\tau)^\alpha\beta} + 1$. This implies that at least $\Omega(n)$ time slots are necessary when applying the power assignment $P(s_i) = f(d(s_i, r_i))$.

However, the power assignment $P(s_i) = \tau^{i\alpha/2}$ allows us to schedule L in one time slot. The interference at r_j caused by the link set $L_j^- = \{l_i | l_i \in L, i < j\}$ with lower indices form a geometric series. The same holds for the links with higher indices $L_j^+ = \{l_i | l_i \in L, i > j\}$. Note that for $i > j$ $d(s_i, r_j) > \nu_i \geq \tau^{i-j}\mu_j$. Hence

$$I_{r_j}(L_j^-) = \sum_{i=1}^{j-1} \frac{P(s_i)}{d(s_i, r_j)^\alpha} < \sum_{i=1}^{j-1} \frac{\tau^{i\alpha/2}}{\mu_j^\alpha} < \frac{\tau^{j\alpha/2}}{\mu_j^\alpha(\tau^{\alpha/2} - 1)},$$

and

$$I_{r_j}(L_j^+) = \sum_{i=j+1}^{\infty} \frac{P(s_i)}{d(s_i, r_j)^\alpha} < \sum_{i=1}^{j-1} \frac{\tau^{i\alpha/2}}{\tau^{\alpha(i-j)}\mu_j^\alpha} < \frac{\tau^{j\alpha/2}}{\mu_j^\alpha(\tau^{\alpha/2} - 1)}.$$

This yields a signal to noise ratio of

$$SINR_{r_j}(L) = \frac{P_{r_j}(s_j)}{I_{r_j}(L_j^-) + I_{r_j}(L_j^+)} > \frac{\frac{\tau^{j\alpha/2}}{\mu_j^\alpha}}{2\frac{\tau^{j\alpha/2}}{\mu_j^\alpha(\tau^{\alpha/2}-1)}} = \beta.$$

This means that all links can transmit simultaneously and hence the scheduling complexity of this instance is constant. $\qquad\square$

In Ref. [22] a related problem called *bidirectional scheduling* is examined as well. To solve this problem, a schedule and a power assignment that allows senders and receivers to change their roles has to be found. The proof presented above can be adapted to the bidirectional scheduling problem for bounded, linear, and superlinear functions f. For sublinear functions, however, such an adaptation is not possible. In fact, there exists a sublinear function, namely the square-root function of the path-loss $f(x) := \sqrt{x^\alpha}$, (cf. Section 4.3.3), which allows to minimize the number of time slots up to a logarithmic factor for multi-slot scheduling bidirectional communication.

4.3.2 Linear Power Assignment

A popular power assignment strategy uses an energy metric and sets the transmission power of the sender of link l_i to a level proportional to $d(s_i, r_i)^\alpha$, i.e., linear in the fading of the signal. As a consequence, the received signal strength at the receiver nodes is in the same order of magnitude for all links. This power assignment function has been studied from various angles, e.g., in Refs. [6, 12, 23, 66, 83].

In Ref. [23], Fanghänel et al. present non-trivial lower bounds for the Multi-slot Scheduling Problem with power control for transmission requests in arbitrary positions. More precisely, they define an interference measure I (we denote it by Υ in this monograph to avoid confusion with the definition of interference) and prove that under linear power assignments in the order of Υ time slots are necessary for the successful transmission of n links. Furthermore, they show that this can be generalized to arbitrary power assignments by losing a factor of $g(L)\log n$. When restricted to the two-dimensional Euclidean space and $\alpha > 2$, $\Upsilon/g(L)$ time slots are necessary. Apart from these lower bounds, Fanghänel et al. propose two randomized algorithms that lead to a schedule using $O(\Upsilon \log n)$ and $O(\Upsilon + \log^2 n)$ time slots with a linear power assignment. Using ILP relaxation and randomized rounding techniques, they show how these results can be extended to multihop scheduling and routing. Subsequently we study their lower bound results.

Consider the situation for a given receiver node r_i under the linear power assignment. The power assigned to its respective sender s_i is $P(s_i) = d(s_i, r_i)^\alpha$, hence the received power from s_i at r_i is $P_{r_i}(s_i) = \frac{d(s_i,r_i)^\alpha}{d(s_i,r_i)^\alpha} = 1$. Consequently, the interference $I_{r_i}(s_j) = \frac{d(s_j,r_j)^\alpha}{d(s_j,r_i)^\alpha}$ caused by concurrent senders s_j cannot amount to more $1/\beta$. If a sender s_j is too close to r_i, i.e., if $d(s_j, r_i) < d(s_j, r_j)$, then the SINR condition for r_i,

$$SINR_{r_i}(s_j) = \frac{P_{r_i}(s_i)}{I_{r_i}(s_j)} = \frac{d(s_j,r_i)^\alpha}{d(s_j,r_j)^\alpha} > \beta$$

cannot be satisfied. In this case, the two links are to be scheduled in separate time slots. In other words, $\min(1, \frac{d(s_j,r_j)^\alpha}{d(s_j,r_i)^\alpha}) = 1$, and an additional time slot is needed. Of all links in L that are either far away or very short, we can tolerate a subset L' satisfying $\sum_{l_j \in L'} \frac{d(s_j,r_j)^\alpha}{d(s_j,r_i)^\alpha} < 1/\beta$. From the perspective of the receiver node, this leads to the following definition of an interference measure.

Definition 4.6 (Measure of Interference [23]). Let L be a set of communication links, where $l_i = (s_i, r_i) \in L$. For a sender or receiver node v, the interference at v caused by L under a linear power assignment is bounded by

$$\Upsilon_v(L) = \sum_{l_j \in L} \min\left(1, \frac{d(s_j, r_j)^\alpha}{d(s_j, v)^\alpha}\right).$$

Using this we define the measure of interference induced by the request set L:

$$\Upsilon = \Upsilon(L) = \max_{l_i \in L}(\max(\Upsilon_{r_i}(L), \Upsilon_{s_i}(L))).$$

Observe that this definition measures the interference at both sender and receiver nodes, even though only receiver node suffers from concurrent transmissions. Nonetheless, the subsequent theorems show that this measure captures the scheduling complexity, i.e., it can be applied to derive lower bounds on the schedule length.

Before stating the theorems let us state three key facts.

Proposition 4.2

(1) Υ is upper bounded by the number of links,
$\Upsilon_v(L) = \sum_{l_j \in L} \min(1, \frac{d(s_j, r_j)^\alpha}{d(s_j, v)^\alpha}) \leq \sum_{l_j \in L} 1 \leq |L|$.

(2) In a feasible schedule, Υ is **constant at the receivers in every time slot**. Otherwise the SINR condition would be violated as β is a constant. More formally, consider a schedule of length T, where L_k is the set of links scheduled in slot k, $1 \leq k \leq T$. For all receiver nodes r_i of the requests in L_k, it must hold that $\Upsilon_{r_i}(L_k) < 1/\beta$ due to the SINR condition.

(3) Υ is **subadditive**, i.e., for two request sets L and L' the inequality $\Upsilon(L \cup L') < \Upsilon(L) + \Upsilon(L')$ is satisfied.

As a consequence, proving a lower bound for the schedule length of $T \in \Omega(\Upsilon f(n))$ for some function f reduces to the following task: Show that the interference Υ of the requests in one time slot of an optimal schedule is bounded by $O(f(N))$. Thanks to the first fact of Observation 4.2, it suffices to prove for all sender nodes in L_k that $\Upsilon_{s_i}(L_k) \in O(f(n))$ for all $k \in 1, \ldots, T$. This property is used in the following proofs.

Theorem 4.7 ([23]). Let T be the minimum schedule length for a set of requests L in a linear power assignment. Then we have

$$T = \Omega(\Upsilon(L)).$$

Proof. Assume we are given an optimal schedule of length T, where L_k is the set of links scheduled in slot k, $1 \leq k \leq T$. Let us consider the time slot k with the highest value $\Upsilon(L_k)$. As mentioned above, it suffices to consider the interference at the senders and prove that $\Upsilon_{s_i}(L_k) \in O(1)$ for all k.

In order to bound $\Upsilon_{s_i}(L_k)$, we need bounds with respect to a receiver close to s_i, since the SINR condition only affects receivers. Let r_c be the sender closest to s_i. We now draw a ball around s_i with radius $d(s_i, r_c)/2$ to separate the senders of the links in L_k into two sets S' and S''. The set S' contains the senders s_j close to s_i, i.e., $d(s_j, s_i) < d(s_i, r_c)/2$, the set S'' the remaining senders of L_k. Let us now bound $\Upsilon_{s_i}(S')$ and $\Upsilon_{s_i}(S'')$.

Due to the triangle inequality we can conclude that for all $s_j \in S'$

$$d(s_j, r_c) \leq d(s_j, s_i) + d(s_i, r_c) \leq \frac{3}{2}d(s_i, r_c). \tag{4.6}$$

Moreover, we have (see Figure 4.6(a) for an illustration)

$$
\begin{aligned}
d(s_i, r_c) \quad &\leq d(s_i, r_j) && \text{since } r_c \text{ is the closest receiver} \\
&\leq d(s_i, s_j) + d(s_j, r_j) && \text{triangle inequality} \\
&\leq \tfrac{1}{2}d(s_i, r_c) + d(s_j, r_j) && \text{definition of } S'
\end{aligned}
$$

This implies

$$d(s_i, r_c) \leq 2d(s_j, r_j). \tag{4.7}$$

Combining Equations (4.6) and (4.7) yields

$$d(s_j, r_j) \geq \frac{1}{3}d(s_j, r_c)$$

We use Equation (4.3.2) to bound $\Upsilon_{s_i}(S')$ with the number of senders in S'.

$$|S'| = \sum_{s_j \in S'} \frac{d(s_j, r_j)^\alpha}{d(s_j, r_j)^\alpha} \leq \sum_{s_j \in S'} \frac{d(s_j, r_j)^\alpha}{\frac{1}{3^\alpha}d(s_j, r_c)^\alpha} \leq \frac{3^\alpha}{\beta},$$

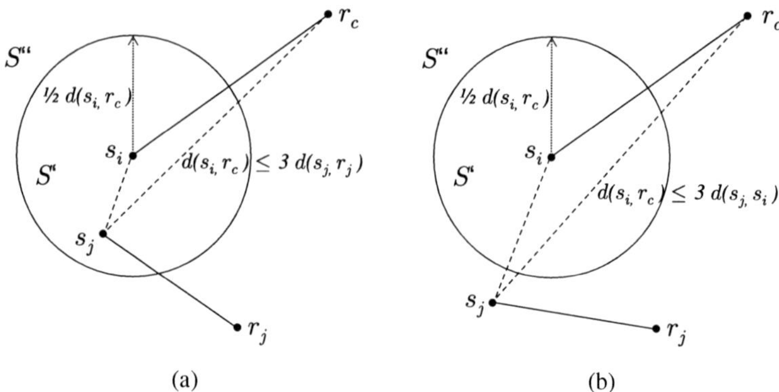

Fig. 4.6 Illustration of scenarios where the sender closest to s_i is (a) in distance less than $1/2d(s_i, r_c)$, (b) in distance greater than $1/2d(s_i, r_c)$.

where the last inequality holds due to the SINR condition for r_c stating that $I_{r_c}(S') < 1/\beta$.

For $s_j \in S''$ we proceed in a similar way (see Figure 4.6(b) for an illustration).

$$d(s_j, r_c) \leq d(s_j, s_i) + d(s_i, r_c) \quad \text{triangle inequality}$$

$$\leq d(s_j, s_i) + 2d(s_j, s_i) \quad \text{definition of } S''$$

$$= 3d(s_j, s_i).$$

This leads to

$$\Upsilon_{s_i}(S'') = \sum_{s_j \in S''} \frac{d(s_j, r_j)^\alpha}{d(s_j, s_i)^\alpha} \leq \sum_{s_j \in S''} \frac{d(s_j, r_j)^\alpha}{\frac{1}{3^\alpha} d(s_j, r_c)^\alpha} \leq \frac{3^\alpha}{\beta},$$

where the last inequality holds due to the SINR condition for r_c $(I_{r_c}(S'') < 1/\beta)$.

Hence we can conclude that

$$\Upsilon_{s_i}(L_k) \leq 1 + \Upsilon_{s_i}(S') + \Upsilon_{s_i}(S'') \leq 1 + \frac{2 \cdot 3^\alpha}{\beta} \in O(1). \qquad \square$$

Thus we have proved that Υ captures the scheduling complexity under linear power assignments. This result can be generalized for arbitrary power assignments. To this end, we need the following lemma.

Lemma 4.8 ([23]). Consider a ball $B_\ell(v)$ of radius ℓ around a center node v. Let L be a feasible set of links of minimum length d, $d \leq \ell$, where all senders are inside this ball $B_\ell(v)$. Then,

$$|L| \leq \frac{1}{\beta}\left(\frac{4\ell}{d}\right)^\alpha + 1.$$

Proof. Let P be a feasible power assignment and let $l_i = (s_i, r_i)$ be a request with minimal power $P(s_i)$. Since the SINR condition is satisfied for the link l_i it must hold that

$$I_{r_i}(L \setminus \{l_i\}) \leq \frac{1}{\beta}\frac{P(s_i)}{d(s_i, r_i)^\alpha}.$$

Because the senders of L are insider the ball $B_\ell(v)$ it holds that

$$I_{r_i}(L \setminus \{l_i\}) \geq (|L| - 1)\frac{P(s_i)}{(2d + 2\ell)^\alpha}.$$

Therefore,

$$(|L| - 1) \leq \frac{1}{\beta}\frac{P(s_i)}{d(s_i,r_i)^\alpha}\frac{(2d_{min} + 2\ell)^\alpha}{P(s_i)} \leq \frac{1}{\beta}\left(\frac{4\ell}{d}\right)^\alpha. \qquad \square$$

Theorem 4.9 ([23]). Let T be the minimum schedule length for a set of requests L with node locations in any metric space using an arbitrary power assignment. Then we have

$$T = \Omega\left(\frac{\Upsilon(L)}{g(L)\log n}\right).$$

Proof. (Sketch) We proceed in a similar way to the proof with linear power assignments. Given an optimal schedule of length T, the requests of time slot t are divided into $g(L)$ classes depending on the link length. Class $C_{t,k}$, contains the requests l_i of time slot t with a length $2^{k-1}d_{\min} \leq d(s_i, r_i) < 2^k d_{\min}$. Using the subadditivity of Υ, it remains to prove that $\Upsilon(C_{t,k}) \in O(\log n)$ for all t, k.

Now we apply Lemma 4.8 for growing balls around the node v with the highest $\Upsilon_v(C_{t,k})$: if the number of senders in the ith ball is 2^i, then the smallest possible radius of the ball is $\ell_i \geq (2^{i-1} - 1)^{1/\alpha}\frac{2^{k-1}}{4}$. Using this knowledge, we can determine that the interference caused by a ring of inner radius ℓ_i and outer radius ℓ_{i+1} is constant. Since there are $O(\log n)$ such rings, this implies that $\Upsilon_v(C_{t,k})$ is upper bounded by $O(\log n)$. $\square$

In the Euclidean space and for $\alpha > 2$, Υ measures the complexity even more closely than in the general case.

Theorem 4.10 [23] Let the instance be located in the Euclidean plane and let $\alpha > 2$. Then we have

$$T = \Omega\left(\frac{\Upsilon(L)}{g(L)}\right),$$

where T denotes the optimal schedule length using any power assignment.

Proof. As in the proof for Theorem 4.3.2, we divide the requests into $g(L)$ classes $C_{t,k}$ containing the links of length $2^{k-1}d_{\min} \le d(s_i, r_i) < 2^k d_{\min}$ scheduled in time slot t. Let v be the node with the highest interference $\Upsilon_v(C_{t,k})$. We now have to show that $\Upsilon(C_{t,k}) \in O(1)$ for all t, k. Let v be the node with the highest interference $\Upsilon_v(C_{t,k})$.

To this end, we partition the plane into an infinite sequence of concentric rings with center v. The inner radius of the ith ring is $i2^{k-1}$ and its width 2^{k-1}. This ring can be covered by i circles of radius 2^{k-1}. Thanks to Lemma 4.8 we know that each of these circles can contain at most $\frac{4^\alpha}{\beta}$ senders. Thus the interference the senders in the ith ring cause at node v is at most $\frac{C}{i^{\alpha-1}}$, for some constant C. Summing up over all i proves that $\Upsilon_v(C_{t,k}) \in O(1)$ due to the fact that $\sum_{i=1}^{\infty} i^{1-\alpha} \le \frac{\alpha-1}{\alpha-2}$. $\square$

These bounds imply that an optimal linear power assignment schedule is an $O(g(L)\log n)$-approximation of the best possible arbitrary power assignment. For the Euclidean space and $\alpha > 2$, the approximation ratio is in $O(g(L)).^2$ Moreover, the bounds can be applied to

---

[2] Observe that this is included in Halldórsson's result that any oblivious power assignment strategy yields an $O(g(L))$-approximation [40].

Algorithm 3 Simple Randomized Linear Power Algorithm [23]

1: **while** packet has not been successfully transmitted **do**
2:    Try transmitting with probability $1/(2(\frac{1}{\beta} - N)\Upsilon)$
3: **end while**

---

construct a simple randomized algorithm for the Multi-slot Scheduling Problem with power control, that produces a schedule of length $O(\Upsilon \log n)$ w.h.p. (see Ref. [23] for a proof). It can be implemented in a distributed way if the nodes know Υ.

Clearly, a main disadvantage of this algorithm is that it does not take the fact that there is less competition for transmission the more time has passed into account. A more sophisticated algorithm increases the probability of transmission over time. Fanghänel et al. [23] propose another algorithm that assigns random delays to all packets and adapts the transmission probability to the current Υ, induced by the request not scheduled yet. This algorithm yields a schedule of length $O(\Upsilon + \log^2 n)$ w.h.p.

This algorithm can be generalized to multihop settings. With linear programming routing problems can be covered as well, see Ref. [23].

In practice, both these algorithms suffer from a significant drawback: they are not applicable for distributed environments, because

---

Algorithm 4 Adaptive Randomized Linear Power Algorithm [23]

1: **while** $\Upsilon_{curr} \geq \log n$ **do**
2:    $J := \Upsilon_{curr}$
3:    **while** $\Upsilon_{curr} \geq J/2$ **do**
4:       **if** packet has not been successfully transmitted **then**
5:          Assign delay d between 1 and $16e(1/\beta - N)J$ independently uniformly at random
6:          Try transmitting after waiting d time slots
7:       **end if**
8:    **end while**
9: **end while**
10: Execute Algorithm 4.3.2

---

assessing the interference of an instance without global knowledge or communication is impossible.

Summarizing this section we have seen that instance-based interference measure proposed in Ref. [23] is similar to the affectedness and affectance defined for uniform power scheduling in Refs. [30, 40]. It coincides with the order of the number of time slots necessary for transmission if a linear power assignment is applied. These ideas, moreover, the authors show how a lower bound for general power assignments can be derived (no non-trivial lower bounds have been known for this case before). Unfortunately, this lower bound is weak for instances with a large link length diversity.

4.3.3 Square Root Assignment

Besides the uniform and the linear power assignment, another oblivious power assignment function has recently received considerable attention. The square root assignment, which sets the power of the sender of a link $l_i = (s_i, r_i)$ to $P(s_i) := \sqrt{d(s_i, r_i)^\alpha}$ is the geometric mean between the uniform and linear power assignment and has been used in Refs. [22] and [40]. Its advantage over the uniform and linear power assignment is most obvious when examining a one-dimensional setting with nested links depicted in Figure 4.4.

More generally, there exists an algorithm using the square root assignment that achieves an $O(\log \log \Lambda \log n)$-approximation for the Multi-slot Scheduling Problem with power control [40].

Theorem 4.11 ([40]). The power control scheduling problem is $O(\log(n) \log(\log(\Delta)))$-approximable with the square root power assignment.

Proof. (Sketch) Remember that with any oblivious power assignment we can compute a constant approximation of the optimal schedule if the lengths of the links differ by a factor of at most 2 (Theorem 4.2). Under the square root assignment, links of widely different lengths can be scheduled together easily as well. Informally, a feasible set containing links of lengths that differ by factor of less than 2 or greater than $n^{2\alpha}$, can be scheduled with $O(\log(\log(\Delta))\max_i T_i)$ time slots, where T_i is the number of time slots necessary to schedule the links in one length class. Thus, the algorithm first partitions L into lengths classes C_k, such that C_k contains the requests l_i with a length $2^{k-1}d_{\min} \le d(s_i, r_i) < 2^k d_{\min}$. Next, it defines sets S_i, consisting of classes of varied lengths:

$$S_i := \bigcup_j C_{i+j\cdot\frac{2}{\alpha}\log n}, \quad \forall i < \frac{2}{\alpha}\log n.$$

For each of these sets S_i, a separate schedule is produced, yielding an approximation ratio of $O(\log(n)\log(\log(\Delta)))$. $\qquad\square$

For a detailed description and analysis we refer the reader to Ref. [40]. Note that depending on the given set of links L, this can be faster or much slower than the bound $O(g(L))$ stated in Theorem 4.3. Very recently, Halldórsson and Mitra have presented an $O(\log(n) + \log(\log(\Delta)))$-approximation algorithm for the single-slot scheduling problem in general metrics in Ref. [41] using mean power. This implies an $O(\log^2(n) + \log(n)\log(\log(\Delta)))$-approximation for the multi-slot scheduling problem in general metrics.

In practical scenarios, nodes constantly swap sender and receiver roles, since they have to acknowledge the reception of messages. Thus, it does not suffice to guarantee unidirectional transmissions. Fanghänel et al. [22] introduce the bidirectional scheduling problem, where a link is scheduled successfully if the role of the sender and receiver of a link can be interchanged and the SINR condition is still met for every link scheduled concurrently. Due to the stronger separation required for bidirectional transmissions, the linear lower bound for unidirectional scheduling does not hold and the authors suggest a randomized algorithm using the square root assignment. They show that this algorithm yields an $O(\log^{4.5+\alpha} n)$ approximation. Shortly thereafter, Halldórsson [40] proved that even an $O(\log n)$-approximation is possible under the square root assignment using a different proof technique based on independent set properties of a graph defined by a scheduling problem instance for fading metrics. In the mean time, Halldórsson and Mitra have presented an $O(1)$-approximation algorithm for the single-slot scheduling problem in general metrics in Ref. [41]. This implies an $O(\log n)$-approximation for the Multi-slot Scheduling Problem in general metrics.

4.4 Arbitrary Power Assignment

In the previous section we investigated oblivious power assignments, where the transmission power of the sender of a communication request only depends on the distance to the receiver. We have seen that even in simple scenarios this strategy fails and leads to unnecessarily long schedules for the Multi-slot Scheduling Problem with power control. Moreover the performance of the algorithms depends on the maximal ratio between the length of two links.

In this section, we study approaches that assign power levels to the senders taking the whole set of communication requests into consideration. More precisely, the power assigned to the senders may depend on the number of other requests and their distribution in the Euclidean plane.

From Section 4.2 we know that we can efficiently determine for a given set of links, what the best power assignment for concurrent transmission is. However, in many cases, it is not possible to schedule all requests in one slot, even with an optimal power assignment, because the maximum achievable SINR might be below the threshold β necessary for a correct decoding.

As a consequence some links have to be postponed to later time slots. In Section 3.1 we have seen that scheduling with uniform power is NP-complete. Although everybody believes that adding power control does not simplify the problem, to the best of our knowledge no NP-hardness proof is known for the scheduling problem in the physical model with power control. All hardness proofs we are aware of have restrictions, e.g., Refs. [17] and [51] assume a bound on the maximum power and Ref. [52]'s proof is based on general metrics.

Nevertheless, a number of algorithms and heuristics have been proposed to solve this problem. Among them are the recursive link removal algorithms that we will examine as a next step. First, we study the measure *disturbance* (introduced in Ref. [65]), which captures the difficulty of a given scheduling task. This enables us to compare algorithms based on their ability to solve "easy" and "difficult" instances.[3]

4.4.1 Disturbance

Since we study arbitrary, possibly worst-case network and request settings, we introduce a formal measure that comprises the intrinsic difficulty of scheduling a given set of communication requests.

For a given set of communication requests L and some constant $\rho \geq 1$, the ρ-*disturbance* is defined as the maximal number of senders (receivers) that are in close physical proximity (depending on the parameter ρ) of any sender (receiver). Consider disks S_i and R_i of radius d_i/ρ around sender s_i and receiver r_i, respectively. Formally, the ρ-*disturbance of a link* l_i is the larger of either the number of senders in S_i or the number of receivers in R_i (see Figure 4.7 for an illustration). The ρ-*disturbance of* L is then the maximum ρ-disturbance of any link $l_i \in L$.

Definition 4.12 ([65]). Given a set of requests L, the ρ-disturbance, denoted as χ_ρ of L is defined as:

$$\chi_\rho := \max_{l_i \in L} \chi_\rho(l_i),$$

where the disturbance $\chi_\rho(l_i)$ for request l_i is the maximum of $|\{r_j \mid d(r_j, r_i) \leq d_i/\rho\}|$ and $|\{s_j \mid d(s_j, s_i) \leq d_i/\rho\}|$.

The *disturbance* of a set of requests indeed captures the fundamental difficulty of the Multi-slot Scheduling Problem with power control for these requests. Solving problem instances with low disturbance efficiently is very important in practice since in realistic networks one always tries to prevent situations with many receivers clustered in the same area.

---

[3] A related measure called I_{in} has been introduced and studied in Ref. [68]. For constant I_{in} the scheduling algorithm presented in Ref. [68] achieves a scheduling complexity of $O(\log^2 n)$. The algorithm proposed in Ref. [65] constructs schedules of length $O(\log^2 n)$ for scenarios with constant disturbance.

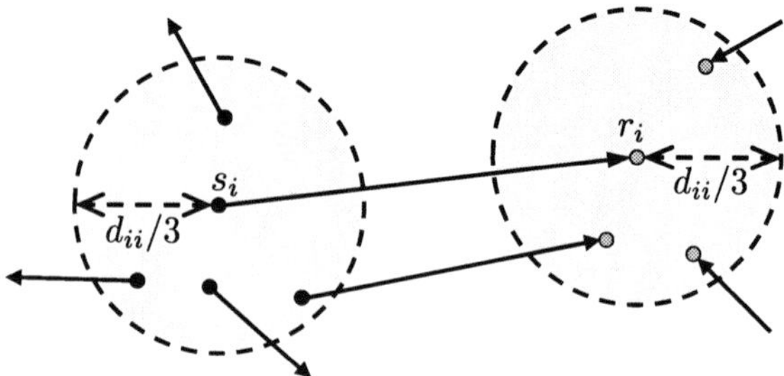

Fig. 4.7 Illustrating example for the disturbance of a communication request l_i. For $\rho = 3$ the disturbance in this case is $\chi_3(l_i) = 4$.

Intuitively, the disturbance of a set of requests in a network characterizes the difficulty of scheduling these requests in a wireless communication environment. Therefore, an efficient scheduling algorithm should be capable of generating short schedules in settings with low disturbance. Unfortunately, all previously known scheduling algorithms may require a linear number of time slots in order to schedule a set of requests even if their ρ-disturbance is as low as 1.

4.4.2   Link Removal Algorithms

In contrast to the intuitive scheduling algorithms with oblivious power assignments, *link removal algorithms* for the Multi-slot Scheduling Problem with power control are much more sophisticated. The heuristics known in the literature are all based on a generic link removal algorithm.

The idea of these algorithms is to postpone the transmission of a link l_k from the set of the links if some condition CON holds, until the minimal $SINR$ level for successful reception is met. Then the optimal power vector is assigned and the procedure is repeated with the remaining links.

We examine the four algorithms SRA, $SMIRA$, $WCRP$, and $LISRA$, which follow the execution of the generic algorithm and differ only in the condition CON.

SRA (Stepwise Removal Algorithm), devised by Zander [86], iteratively removes the link with the largest row or column sum of Z, since these sums provide a bound on the maximal eigenvalue, until

---

Algorithm 5 Generic Link Removal Algorithm

1: Time slot $t := 1$;

2: **while** there are links to schedule **do**

3:     Compute $SINR^*$ and $\mathbf{P}^*$ from Z;

4:     **while** $SINR^* \leq \beta$ **do**

5:         Remove links l_k for which CON is satisfied;

6:         Compute $SINR^*$ and $\mathbf{P}^*$ from new Z;

7:     **end while**

8:     Schedule the links of Z in time slot t and assign $\mathbf{P}^*$;

9:     Time slot $t := t + 1$;

10:     Compute new Z for unscheduled links;

11: **end while**

---

the required $SINR$ level is met.

$$CON:\ \max\left\{\sum_j Z_{kj}, \sum_j Z_{jk}\right\} \text{ is maximimal for } k.$$

SMIRA (Stepwise Maximum Interference Removal Algorithm), by Lee et al. [60], excludes links which cause or receive the most interference when power is assigned optimally, taking the normalized link gain matrix Z and the corresponding optimal power vector into account.

$$CON:\ \max\left\{\sum_{j\neq k} P_j Z_{kj}, P_k \sum_{j\neq k} Z_{jk}\right\} \text{ is maximimal for } k.$$

Lee et al. suggest versions of this algorithm considering only $\max_k(\sum_{j\neq k} P_j Z_{kj})$ or $\max_k(P_k \sum_{j\neq k} Z_{jk})$ in the condition and demonstrate with simulations, that they perform worse than SMIRA. Our analysis can be adapted easily to these cases with the same complexity result.

WCRP is a (distributed) algorithm presented in Ref. [82]. When adapted to our model, it first computes for each row i the value $MIMSR$ (maximum interference to minimum signal ratio), defined by

$$MIMSR(i) = \max\left\{\frac{\beta G(i,j)}{G(i,i)} | j \neq i \wedge j \text{ not scheduled}\right\},$$

and removes links with MIMSR above a threshold ζ. We present here a simplified and centralized version, which produces schedules of at most the same length as the original algorithm.

$$CON: MIMSR(k) > \zeta.$$

LISRA (Limited Information Stepwise Removal Algorithm), described in Ref. [87], postpones the transmission of the links with the lowest $SINR$ when all senders transmit with equal power, to increase the probability for the remaining links to reach the $SINR$ threshold.[4] To generate schedules with LISRA we replace Step 5 of the generic with

5a: set $\mathbf{P} = 1$ and compute $SINR$ for remaining link set S;
5b: remove links γ_k for which $\min_i SINR_i(S) = SINR_k(S)$;

$$CON: SINR_k(S) \text{ is minimal for } k.$$

These algorithms have all been tested in situations with nodes distributed uniformly at random. No worst-case analysis has been done and the authors do not give any guarantees on their behavior. To prove this point we construct an example where the schedules these algorithms produce are extremely long.

Consider Example 4.4 of n nested communication requests. This scenario is depicted in Figure 4.4. It has been used in Section 4.3 to show that the uniform and the linear power assignment strategy can lead to arbitrarily bad performance. All sender and receiver nodes are situated on a straight line and the distance between sender and receiver nodes is increasing exponentially. We set $x = 2$, $\alpha = 3$, the noise level $N = 0$ and the minimum $SINR$ necessary for successful transmission to $\beta = 2$. For this situation all the algorithms described above perform poorly, namely they schedule each link individually and require $\Omega(n)$ time slots, even though a constant number of time slots is proven to be sufficient in Section 4.3. Because the 3-disturbance of the above scenario is $\chi_3 = 1$,

---

[4] In its original version, step 3 contains the execution of an iterative distributed algorithm based on locally available information. The number of rounds is fixed beforehand, hence the quality of the results depends on the convergence speed of the algorithm. As we are most interested in the maximal length of the schedules LISRA produces, we replace the algorithm in step 3 by a (centralized) eigenvalue decomposition.

our example demonstrates that these algorithms exhibit severe worst-case problems even in networks with low disturbance.

Theorem 4.13 ([65]). SRA, LISRA, SMIRA, and WCRP produce a schedule of length $\Omega(n)$ for Example 4.4 in which the 3-disturbance χ_3 is 1.

Proof. Starting with SRA, we prove the claim for each algorithm individually.

SRA: As we cannot schedule all links in the same slot, we compute the column and row sums of Z to decide which links we postpone to subsequent time slots. The sum for row i is:

$$R_i = \sum_{j=1}^{n} z(i,j) = \sum_{j=1}^{n} \left(\frac{2^{i+1}}{2^j + 2^i} \right)^{\alpha},$$

which is maximal when $i = n$. Analogously the sum for column i is:

$$C_i = \sum_{j=1}^{n} z(j,i) = \sum_{j=1}^{n} \left(\frac{2^{j+1}}{2^j + 2^i} \right)^{\alpha}.$$

This sum reaches its maximum when $i = 1$, since i only appears in the denominator. Hence we have to determine $\max\{R_n, C_1\}$.

The summands of C_1 grow with j whereas the summands of R_n decrease. As a consequence we can simplify the analysis by comparing $\frac{2^{n+1}}{2^{n-j+1}+2^n}$ with $\frac{2^{j+1}}{2^j+2}$.

$$\frac{2^{n+1}}{2^{n-j+1} + 2^n} = \frac{2^j}{1 + 2^{j-1}} - \frac{2^{j+1}}{2 + 2^j} \quad \forall 0 < j \le n.$$

Hence we know that the largest row sum is equal to the largest column row, which causes either the shortest or the longest link to be removed from the set of links to schedule in the next time slot. Without loss of generality we assume that we postpone the transmission of the shortest link.

Without the first link we have to deal with almost the same situation, the only difference is the fact that the sums start with $j = 2$ instead of 1. Again we remove the shortest link. This game continues until only one link is left, since two links next to each other cannot be scheduled simultaneously.

Lemma 4.14 ([65]). Two links l_i and l_{i+1} cannot be scheduled simultaneously.

Proof. Let $l_i = (-2^i, 2^i), L_j = (-2^j, 2^j)$. We compute

$$
Z = \begin{pmatrix} 1 & \left(\dfrac{2^{i+1}}{2^j + 2^i}\right)^\alpha \\[2ex] \left(\dfrac{2^{j+1}}{2^j + 2^i}\right)^\alpha & 1 \end{pmatrix}
$$

and set $j = i + 1$. Now the larger eigenvalue is:

$$
\begin{aligned}
\lambda^* &= 1/2 \left(z_{1,1} + z_{2,2} + \sqrt{4z_{1,2}z_{2,1} + (z_{1,1} - z_{2,2})^2} \right) \\
&= 1/2 \left(1 + 1 + \left(\sqrt{4 \cdot 2^{i+j+2}/(2^i + 2^j)^2} \right)^\alpha \right) \\
&\overset{j=i+1}{=} 1 + \left(\frac{\sqrt{2^{2i+3}}}{2^i + 2^{i+1}} \right)^\alpha = 1 + \left(\frac{\sqrt{8}}{3} \right)^\alpha > 1.83.
\end{aligned}
$$

Consequently $SINR^* = \frac{1}{\lambda^* - 1} < 1.19$, implying that the links l_i and l_{i+1} cannot be transmitted simultaneously. $\qquad\square$

We can derive from the above, that SRA schedules all links individually, i.e., the length of the schedule is $\Omega(n)$.

SMIRA: The transmission of link l_i is postponed if either the interference received and the interference caused by link l_i is above a certain threshold. As the receiving node of link 1 suffers from the highest level of interference we remove it. This situation occurs again in the next time slot, hence each link is scheduled individually, leading to a complexity of $\Omega(n)$.

WCRP: We compute the MIMSR value for each link i.

$$
MIMSR(i) = \max_j \frac{\beta \cdot G(i,j)}{L \cdot G(i,i)} = \beta \cdot \max_i \left(\frac{2^{i+1}}{2^i + 2^j} \right)^\alpha.
$$

As $MIMSR(i)$ cannot exceed $\beta 2^\alpha$, we define $\zeta = 10$. Hence all links apart from the three shortest links are removed. Let us assume for simplicity that those can be scheduled in one slot. If we repeat this step, again the three shortest links remain and we can conclude that this method produces a schedule of length $\lceil n/3 \rceil \in \Omega(n)$.

LISRA: The same holds for LISRA, although with a slightly different reasoning. LISRA iteratively removes the link which achieves the lowest $SINR$ with equal power distribution until β is reached. In our example, the link to be postponed will always be the longest link. As we have seen above, two neighboring links cannot be scheduled in the same time slot, hence LISRA also needs $\Omega(n)$ slots. $\square$

All four algorithms produce a schedule of length $\Omega(n)$ for this example. However, as we have seen in Section 4.3.3, it is possible to schedule all links in one time slot if $x > 2^{2+2/\alpha}\beta^{2/\alpha}$. Since $x := 2^4$ satisfies this condition, we can schedule every fourth link starting with the shortest link in one time slot. By repeating the same with the remaining links, we can construct a schedule of length 4. Thus, we have shown that link removal algorithms cannot guarantee short schedules, even when the disturbance is low. In the next section, we examine an algorithm that bases scheduling decision on the geometry of the instance at hand.

4.4.3   Low-Disturbance Scheduling Algorithm

In this section, we present the scheduling algorithm *Low-Disturbance Scheduling (LDS) algorithm* proposed in [65]. It gives provable performance guarantees for the Multi-slot Scheduling Problem with power control even in worst-case networks. In particular, given a network and a set of communication requests, LDS algorithm computes a schedule whose length is within a polylogarithmic factor of the network's disturbance.

Description

The algorithm consists of three parts: a pre-processing step, the main scheduling-loop, and a test-subroutine that determines whether a link is to be scheduled in a given time slot.

The purpose of the pre-processing phase is to assign two values $\tau(i)$ and $\gamma(i)$ to every request l_i. The value $\gamma(i)$ is an integer value between 1 and $\lceil \log(3n\beta) + \rho \log \alpha \rceil$. The idea is that only requests with the same $\gamma(i)$ values are considered for scheduling in the same iteration of the main scheduling-loop (Lines 2 and 3 of the main scheduling-loop). The second assigned value, $\tau(i)$, further partitions the requests. In particular, it holds that the length of all requests that have the same $\gamma(i)$ and $\tau(i)$ differ by at most a factor of 2. On the other hand, one can show that if two requests l_i and l_j satisfy $\tau(i) < \tau(j)$, then the length of l_i, d_i, is at least by a factor $\frac{1}{2}(3n\beta\rho^\alpha)^{\tau(j)-\tau(i)}$ longer than d_j. Generally speaking, the assignment of $\tau(i)$ ensures that the smaller the value $\tau(i)$ assigned to a requests l_i, the longer the corresponding communication link, and vice versa.

In summary, the pre-processing phase partitions the set of requests in such a way that two requests l_i and l_j that are assigned the same $\gamma(i)$ have either almost equal length (if, $\tau(i) = \tau(j)$) or very different length. This partition will turn out to be crucial in the actual scheduling process, which takes part in the subsequent main scheduling-loop.

Each for-loop iteration of the main scheduling-loop schedules the set of requests having the same $\gamma(i)$ values, denoted by $\mathcal{F}_k$. As long as not all requests of $\mathcal{F}_k$ have been successfully scheduled, the algorithm considers the remaining requests in $\mathcal{F}_k$ in decreasing order of their length d_i. Specifically, the algorithm checks for each request whether it can safely be scheduled alongside the longer links that have already been selected. If a request is chosen to be scheduled in time slot t, it is added to L_t, otherwise it remains in $\mathcal{F}_k$.

The decision whether a request l_i is selected for scheduling or not takes place in the **allowed$(\mathbf{l_i}, \mathbf{L_t})$** subroutine. For each (longer) request $l_j \in L_t$ that has already been chosen to be scheduled in time slot t, the subroutine checks three conditions. Only if none of them is violated, l_i is added to L_t. Notice, however, that the selection-criteria are significantly more complex than the "reuse-distance" argument that has been used in other work (e.g., Ref. [19]). In particular, the second criterion states that l_i is scheduled only if for all longer requests $l_j \in L_t$, it holds that $d_i \cdot (3n\beta\rho^\alpha)^{\frac{\tau(i)-\tau(j)+1}{\alpha}} > d(s_i, r_j)$ if $\tau(i) > \tau(j)$. That is, the distance that must be maintained between the sender s_i of l_i and the receiver of r_j

of some $l_j \in L_t$ depends on the relative values of $\tau(i)$ and $\tau(j)$ assigned in the pre-processing phase.

The links in L_t are assigned a power level proportional to the ambient noise N, the length of the link to the power of α and factor exponential in $\tau(i)$ in Line 10.

The definition of the three selection-criteria guarantees that all simultaneously transmitted requests in a single time slot are received successfully by the intended receivers. Additionally, the subsequent analysis section shows that all requests can be scheduled efficiently even in worst-case networks.

This algorithm provably schedules every set of requests efficiently even in worst-case networks provided that the ρ-disturbance of the requests is small. As we have demonstrated above, this distinguishes the LDS algorithm from scheduling algorithms with oblivious power assignments and link removal algorithms, that may perform badly even if the disturbance is small.

Theorem 4.15 ([65]). The number of time slots required by Algorithm 6 to successfully schedule all requests $l_i \in L$ is at most $O(\chi_\rho \rho^2 \log n \cdot (\log n + \rho))$.

Algorithm 6 The LDS Algorithm for requests L [65]

Pre-processing phase:

1:   $\tau\text{cur} := 1$; $\gamma\text{cur} := 1$;   last $:= d_1$;

2:   Consider all requests $l_i \in L$ in decreasing order of d_i:

3:   **for** for each $l_i \in L$ **do**

4:     **if** last$/d_i \geq 2$ **then**

5:       **if** $\gamma\text{cur} < \lceil \log(3n\beta) + \rho \log \alpha \rceil$ **then**

6:         $\gamma\text{cur} := \gamma\text{cur} + 1$;

7:       **else**

8:         $\gamma\text{cur} := 1$; $\tau\text{cur} := \tau\text{cur} + 1$;

9:       **end if**

10:       last $:= d_i$;

11:     **end if**

12:     $\gamma(i) := \gamma\text{cur}$; $\tau(i) := \tau\text{cur}$;

13: **end for**

Main scheduling-loop:

1: $t := 1;\ \nu := 4N$;
2: **for** $k = 1$ **to** $\lceil \log(3n\beta) + \rho \log \alpha \rceil$ **do**
3:     Let $\mathcal{F}_k$ be the set of all requests l_i with $\gamma(i) = k$.
4:     **while** not all requests in $\mathcal{F}_k$ have been scheduled **do**
5:         $L_t := \emptyset$;
6:         Consider all $l_i \in \mathcal{F}_k$ in decreasing order of d_i:
7:         **if** $allowed(l_i, L_t)$ **then**
8:             $L_t := L_t \cup \{l_i\};\quad \mathcal{F}_k := \mathcal{F}_k \setminus \{l_i\}$
9:         **end if**
10:         Schedule all $l_i \in E_t$ in time slot t, assigning s_i
              a transmission power of $P_i := \nu \cdot d_i^\alpha \cdot (3n\beta\rho^\alpha)^{\tau(i)}$;
11:         $t := t + 1$;
12:     **end while**
13: **end for**

allowed(l_i, L_t)

1: Define constant μ such that $\mu := 4 \sqrt[\alpha]{\frac{120\beta(\alpha-1)}{\alpha-2}}$;
2: **for** $l_j \in L_t$ **do**
3:     $\delta_{ij} := \tau(i) - \tau(j)$;
4:     **if** $\tau(i) = \tau(j)$ and $\mu \cdot d_i > d(s_i, s_j)$
       or $\tau(i) > \tau(j)$ and $d_i \cdot (3n\beta\rho^\alpha)^{\frac{\delta_{ij}+1}{\alpha}} > d(s_i, r_j)$
       or $\tau(i) > \tau(j)$ and $d_j/\rho > d(s_j, r_i)$ **then**
5:         return false
6:     **end if**
7: **end for**
8: **return true**

Proof. (Sketch, see Ref. [65] for details). The interference of concurrent senders is bounded for links of the same length and for links of different length. This is achieved by determining the maximum number of senders and there power level in a sequence in concentric rings around receiver nodes. This guarantees correctness in the sense that in every time slot all messages are actually received successfully. It now remains to show that the schedule is short and includes all requests. For this reason, the number of time slots required to schedule all requests that have the same $\gamma(i)$ value is bounded. Thus, an upper bound on the amount of time used for one iteration of the for-loop in the main scheduling-loop is reached. Then the delay caused by the reuse criteria in the subroutine **allowed**$(\mathbf{l_i}, \mathbf{L_t})$ is determined, which completes the proof. $\qquad\square$

Let us now examine the schedule the LDS algorithm creates for the instance of Example 4.4 with n nested communication requests. Recall that we used this example to illustrate the inefficiency of previous algorithms with $x = 2, \alpha = 3, N = 0$ and the minimum *SINR* necessary for successful transmission set to $\beta = 2$. The 3-disturbance χ_3 of this setting is 1. Consequently, we obtain a schedule of length $O(\log^2 n)$ by plugging in the value $\rho = 3$ into the bound of Theorem 4.15. Notice that this is *exponentially shorter* than the schedules generated by any uniform or linear power assignment algorithm as well as any of the known link removal heuristics.

Corollary 4.16 ([65]). For $\rho = 3$, the LDS scheduling algorithm produces a schedule of length $O(\log^2 n)$ for Example 4.4 with $x = 2$, $\alpha = 3$, $N = 0$ and the minimum *SINR* necessary for successful transmission set to $\beta = 2$.

The LDS algorithm thus significantly outperforms link removal scheduling strategies in worst-case scenarios. By employing a different power assignment scheme and reuse distance criterion, this algorithm achieves a provably efficient performance in any network and request setting that features low disturbance. Nonetheless, an oblivious power assignment with constant scheduling complexity exists for this scenario, as we have seen earlier. This assignment has of course the drawback, that it can lead to arbitrarily bad performance in other scenarios, as illustrated in Theorem 4.5. While the LDS algorithm thus remedies some of the drawbacks of previous approaches, it is completely centralized and hence suitable to be employed in static networks with known traffic patterns only. Whether a distributed algorithm working in a manner similar to this algorithm exists, is an open question. Ideally, such a distributed worst-case efficient scheduling algorithm could lead to improved MAC-layer solutions, as combined power control and scheduling are crucial to a theoretical understanding of media access control problems.

4.4.4 Approximation Algorithms

Dinitz [16] and Dinitz and Andrews [17] used game theoretic approaches to construct distributed algorithms for scheduling with power control. Their results have been improved upon by Ásgeirsson and Mitra [2] very recently to a distributed $O(\log \Delta)$-approximation algorithm for one-slot scheduling.

Very recently, Kesselheim [53] described the first algorithm with a non-trivial approximation ratio independent of the length diversity, the aspect ratio or the link length ratio. For the one-slot scheduling problem, this algorithm achieves a constant approximation for fading metrics (see Ref. [40] for a definition, e.g., this includes nodes in the Euclidean plane with $\alpha > 2$) and an $O(\log n)$-approximation for arbitrary metrics. The centralized algorithm uses a greedy strategy similar to the algorithm for uniform power treated in Section 3.3. The links are processed one after each other, sorted by their length. If the current link satisfies some feasibility criterion together with the links chosen already, it is added to set to schedule, otherwise it is discarded. In the uniform case, the affectance of the current link set is considered. In Ref. [53], the current link l_i is accepted if

$$\sum_{l_j \in S} \frac{d(s_j, r_j)^\alpha}{d(s_j, r_i)^\alpha} + \frac{d(s_j, r_j)^\alpha}{d(s_i, r_j)^\alpha} \leq \frac{1}{2 \cdot 3^\alpha (4\beta + 2)},$$

where S is the set of links selected already. The transmission power assigned is proportional to the minimum transmission power needed for successful simultaneous transmissions with longer links.

By applying this algorithm recursively, we obtain an $O(\log n)$-approximation algorithm for the Multi-slot Scheduling Problem in fading metrics and an $O(\log^2 n)$-approximation for arbitrary metrics.

The algorithm can be extended to solve other problems like the weighted one-slot scheduling problem, the k-channel scheduling problem, and multihop scheduling. The approximation ratios are $O(\log n)$, $O(\sqrt{k}\log n)$, $O(\log^2 n)$ for fading metrics and $O(\log^2 n)$, $O(\sqrt{k}\log^2 n)$, $O(\log^3 n)$ for arbitrary metrics.

4.4.5 Exact Algorithms

So far, all algorithms presented in this chapter computed a schedule and power assignment efficiently, i.e., in polynomial time with regard to the input size. Even though the analysis of some of the discussed algorithms gives a provable guarantee on the approximation ratio, the resulting schedules are not optimal. Hua and Lau [46] propose algorithms that compute a minimal length schedule. However, these algorithms require exponential time and space.

4.5 Outlook

In this chapter, we have seen that power control can increase the number of concurrent transmissions. In some settings, n senders can transmit simultaneously when adjusting their transmission power level, whereas with uniform power every single transmission needs its own time slot. Solving an Eigenvalue problem yields the optimal power assignment for a given set of links, i.e., one can determine efficiently, whether a set of links can be scheduled in one time slot.

We have considered oblivious power assignments, where the transmission power only depends on the link length. For these assignments, we gave lower and upper bounds on their performance compared to arbitrary power assignments. The greatest advantage of such an assignment is the fact that it can be used to construct truly distributed algorithms, since the power level is not based on global knowledge such as the number of links to be scheduled, the position of other senders, etc. The opposite approach is adopted for link removal, the LDS algorithm and most approximation algorithms. They take all communication requests, i.e., the positions of all sender and receiver nodes, into account for the construction of a short schedule and power assignment. Their most obvious drawback is the fact that they can only be used in static scenarios, where the overhead of computing a centralized solution is acceptable. In addition, the resulting schedule is not necessarily optimal. Thus, the quest for efficient scheduling and power control algorithm is not finished yet. Moreover, a non-trivial lower bound, where despite power control more than a constant number of slots are necessary, is yet to be found. I.e., to the best of our knowledge no instance has been constructed where even with power control the scheduling complexity exceeds a constant.

From a more practical perspective, it can be argued that the network topologies and request sequences found in real-world applications may not have an explicit worst-case structure. We hope, however, that the theoretical insights gained from the worst-case analysis will ultimately lead to an increase in bandwidth and capacity beyond heuristics in real networks. Further investigation in this direction will lead to useful results in areas such as wireless mesh networks, sensor networks, or even cellular networks.

5

Related Problems

The focus of this survey is scheduling a set of communication requests in the physical interference model. In this chapter, we discuss a few results of other problems in this interference model. We believe that many of the papers written for protocol models should be reconsidered in the physical model. Similarly, many of the results that hold on special-case topologies in the physical model may be re-considered and generalized in an algorithmic way.

5.1 Topology Control and Connectivity

In order to save energy and thus extend the lifetime of a network, topology control is applied. The nodes of a network coordinate their transmission power yielding a network topology with desirable properties. Thus the main goal is to reduce the number of active links and yet guarantee a connected network. In addition to connectivity, other properties such as low node degrees, sparseness, or planarity might be required.

Using results from percolation theory, Gupta and Kumar [37] investigate the critical power level that is necessary for a randomly deployed wireless network to become connected under the assumption that all nodes transmit at the same power level. Ever since, much research effort has been directed toward studying asymptotic connectivity requirements in randomly distributed wireless networks, e.g., [18, 84]. What these papers do not consider, however, is the complexity of actually scheduling the communication links that form the connected network.

Moscibroda et al. [66] study the problem of scheduling a strongly connected set of links given n nodes and their positions in the plane. They propose a power control and scheduling algorithm that can successfully schedule such a set is in $O(\log^4 n)$ time slots. This result holds for arbitrary worst-case networks. The power level assigned to senders of short links is higher than actually required to reach the receiver. Yet the transmission power still increases monotonically with the length of the links. Such a sophisticated power assignment strategy is necessary, because uniform and linear power assignments lead to schedules of length n in worst-case networks. However, these lower bounds are based on networks in which some communication links are exponentially longer than others. Moscibroda et al. improve on this work in Ref. [68] by proving that the minimum number of time slots to schedule an arbitrary topology is proportional to the squared logarithm of the number of network nodes times a previously defined static interference measure. Moreover, they demonstrate that topologies that require bidirectional (symmetric) links may lead to significantly higher bounds on the number of time slots necessary. The main result of Ref. [64] (see Section 5.3) implies that the scheduling complexity of connectivity is in $O(\log^2 n)$. Recently, Kowalski and Rokicki [55] devised an algorithm that uses $\mathcal{O}(\log n)$ colors only. They achieve this with a general reduction between a different interference model and by applying an algorithm proposed in Ref. [24]. The $\Omega(\log n)$ lower bound for Minimum Interference Sink Trees [24] can be adapted to the scheduling complexity of connectivity. This implies that the $O(\log n)$ approximation in Ref. [55] is asymptotically optimal.

Even though the number of time slots necessary to guarantee connectivity under uniform power can be in the order of the number of nodes, some devices are simply not able to change their transmission power. Avin et al. examined the complexity of connectivity of a uniform power network in Ref. [4]. More precisely, they analyze grid networks and nodes distributed uniformly at random.

While Refs. [4, 66, 68] strive to minimize the schedule length for the topology of a given set of nodes, a complimentary approach is adopted in Ref. [26]. Gao et al. focus on constructing a topology and a power assignment that maximizes the network capacity. They apply a recursive algorithm that converges to an optimal point for their objective function. They validate the algorithm with simulations and show that it outperforms existing topology control algorithms.

5.2 Online Algorithms

In addition to static scheduling problems, a dynamic version where communication request arrive dispersed over time has been studied as well. In Ref. [40] an $O(\log \Delta)$-competitive algorithm is proposed. Erlebach and Grant [20] extend this result for multicast requests. A multicast request is a set of links with a common sender. One transmission suffices for such a request if the received signal strength exceeds the required SINR threshold at all receivers. Moreover, they give a lower bound of $\Omega(\log \Delta)$ for the competitive ratio of every deterministic online algorithm with arbitrary power assignments. This bound even holds for the unicast case. For an extended scenario with communication request in the Euclidean space of dimension d and a duration in $[1, \Gamma]$ a lower bound for the competitive ratio of $\Omega(\Gamma \Delta^{d/2})$ and a near-optimal upper bound of $\Omega(\Gamma \Delta^{d/2+\epsilon})$, for any constant $\epsilon > 0$ is presented in Ref. [21] for deterministic algorithms. The authors also devise a randomized $O(\log \Gamma \log \Delta)$-competitive algorithm and show how to generalize their ideas to k channels.

5.3 Data Gathering

Much research for wireless networks addresses wireless sensor nodes, devices capable of sensing physical phenomena equipped with a radio communication system. The most important task in a wireless sensor network is to collect the sensed data in a fast and energy-efficient way. Hence the data has to be gathered at an information sink by transmitting messages in a (possibly multihop) way toward the sink. The performance of sensor networks is thus characterized by the rate at which information can be aggregated to the sink.

If all the sensors are within communication range of the sink, they can sequentially transmit their data to the sink. Since there is no interference, the rate scales in $O(1/n)$, where n is the number of nodes. Surprisingly, assuming the protocol model, there exist node distributions where transmitting sequentially is indeed the best one can do, hence in the protocol model the best possible rate is $O(1/n)$ [66].

In randomly deployed networks, i.e., scenarios of nodes distributed uniformly at random, a rate of $O(1/\log n)$ is possible [58] in the protocol model. When the function to be computed on the data exhibits certain properties, this result can be improved, e.g., so-called type-threshold functions can be computed at a rate of $O(1/\log\log n)$ using the block-coding technique. Moscibroda [64] studies the scaling laws of the achievable rate in arbitrarily deployed sensor networks. He proves that a rate of $\Omega(1/\log^2 n)$ is achievable, even in worst-case scenarios. This has been improved recently by Kowalski and Rokicki [55] to $\Omega(1/\log n)$ (see Section 5.1).

5.4   Distributed Protocols

To the best of our knowledge, there is not much work on distributed algorithms in the physical interference model.

In Ref. [75] a distributed algorithm for establishing a dominating set in the physical model is presented. The proposed protocol is randomized, makes extensive use of physical carrier sensing, and converges to a dominating set within a logarithmic number of communication rounds, w.h.p., achieving a dominating set with $O(1)$ approximation bound.

In Ref. [59] an algorithm is proposed to emulate a Unit Disk Graph (UDG)-like structure in the physical model and a network where nodes are distributed uniformly at random on the plane. It is shown that it is possible to emulate a UDG with radius $1/\sqrt{n\ln n}$, that satisfies the SINR constraints, over any set of n wireless nodes uniformly distributed in the unit square, with an $O(\ln^3 n)$ time and power stretch factor.

In Ref. [31], two distributed algorithms for the problem of local broadcasting are proposed. One is a very simple Aloha-like algorithm that is based on the assumption that each node knows the number of its neighbors; the other is more involved and makes no assumptions about topology knowledge. It is shown that, if the transmission probabilities of nodes are carefully set, the global nature of interference in the physical interference model can be separated into "close-in" and "far-away" regions, which allows the analysis to proceed similarly to analysis in graph-based models, such as the protocol model.

5.5   Cross-Layer Protocols

In Ref. [12] the joint cross-layer problem of scheduling, power control and routing is studied. The authors apply random delays to solve the scheduling problem, linear power assignment to solve the power-control problem, and linear programming rounding to solve the routing part of the problem. Let $\mathcal{S}_{OPT}(p_{\min}, (1 - \epsilon)p_{\max})$ denote the optimal latency of minimum length possible for power levels chosen from the range $[p_{\min}, (1 - \epsilon)p_{\max}]$, for any given parameter $\epsilon > 0$. For Λ the aspect ratio as in Definition 2.3, an algorithm with approximation guarantee of $O(\log^2 n \log^3 \Lambda \cdot \mathcal{S}_{OPT}(p_{\min}, (1 - \epsilon)p_{\max})/\log\log n))$ is devised. This work has been further generalized in Refs. [23, 53], where improved upper and lower bounds for the cross-layer problem are presented.

5.6   Network Coding

Network coding is a technique that extends the traditional definition of routing by allowing routers to not just forward copies of received messages, but to mix the bits from different packets before forwarding them. The topic has received a lot of attention in the research community, starting with the pioneering work of Ahlswede et al. [1], where the authors prove that full capacity (i.e., the maximum flow or minimum cut between a source and a receiver) can be achieved in a graph where one source multicasts information to other nodes in a multihop fashion and any node in the network is allowed to encode all its received data before passing it on.

Network coding in the physical layer, or *analog network coding*, is similar in spirit to digital network coding. However, it operates on the raw analog signal, instead of first decoding and then mixing packets in a bitwise manner. Some techniques, such as *cochannel signal separation*, explore differences in the characteristics of interfered signals, such as the signal's strength, to decode several signals simultaneously [43, 44]. Other analog coding techniques exploit the fact that, in a wireless network, often a receiver has prior knowledge about some packets intended to other nodes, by having overheard or forwarded them earlier. This situation has been extensively studied in the context of 2-way relay channel [50, 57, 72, 73].

In Ref. [88] an algorithm for separating two physical-layer signals using higher level information is proposed. The approach is not directly implementable in practice, though, because of several assumptions that the authors make, e.g., they assume that the interfering signals are synchronized at the symbol boundaries and that both signals have undergone the same attenuation when arriving at the router. These problems are overcome in Ref. [49], where analog network coding is made more practical. The authors propose a communication scheme, where pairs of nodes that wish to exchange packets through a relay node are encouraged to transmit simultaneously. The relay node, without decoding the collided signal, amplifies and forwards it. The signal resulting from a collision is the sum of the two colliding signals after incurring attenuation and phase and time shifts. Since the receiver often knows the content of the packet that interfered with the packet it wants, it can cancel the signal corresponding to that known packet after correcting for channel distortion, and the receiver is left with the signal of the packet it wants, which it decodes using standard methods. Thus the destination nodes can extract the packet intended for them by filtering out their own contribution from the mixed signal.

In Ref. [33] the combined problem of analog network coding and link scheduling in the physical interference model is studied. Two definitions of analog network coding are introduced: one definition that uses cochannel signal separation to decode several messages simultaneously, and another definition that is based on the "amplify and forward" approach of Katti et al. [49]. The authors show that, in spite of the ability to decode several signals simultaneously, the scheduling problem remains NP-complete in both models, and propose algorithms to schedule an arbitrary set of links using cochannel signal separation.

5.7 Capacity of Random Networks

Throughput capacity of randomly deployed wireless networks has been studied intensely using an information theoretic approach. In their seminal work [38], Gupta and Kumar provide upper and lower bounds on the capacity of networks with two kinds of topology: one where nodes are distributed uniformly at random in a disk of unit area, and one where nodes are "optimally" distributed on a regular grid lattice. In the former case, the authors show that if each node is capable of transmitting W bits per second, the per node capacity of the network with n nodes is $\Theta(W/\sqrt{n\log n})$. In the "optimum" topology and traffic pattern, the capacity is $\Theta(W/\sqrt{n})$. These results hold in both the protocol and the physical interference models and hold a rather pessimistic character, since they essentially state that large networks cannot achieve high throughput.

Using similar techniques, the capacity of random networks has been further analyzed in many different contexts, such as networks with an overlaid infrastructure [56], multichannel networks [8, 9], MIMO [13], multi-user cooperation [70], use of relays [28], multicast [61], data gathering [58], and cognitive networks [81]. A thorough overview of interference models and capacity results from an information theoretic perspective can be found in Refs. [39] and [85].

6

Alternative Interference Models

In this final section, we study alternative interference models. Section 6.1 discusses graph-based connectivity and interference models. In Section 6.2, we discuss a few other physical models in which it might also be interesting to design and analyze algorithms.

6.1 Graph-Based Models

As already mentioned in the introduction, a popular way to model wireless networks are *graphs*. A graph model usually consists of a connectivity graph and possibly also of an interference graph. In both graphs, the set of vertices represents the devices, and a successful transmission occurs when the sender–receiver pair is connected in the connectivity graph and no other concurrently scheduled sender–receiver pair inflicts a conflict in the interference graph. As a consequence, graph-based scheduling algorithms usually employ some sort of matching or coloring strategy.

There is a vast and rich body of literature on graph-based models, and we are going to mention just a few of them here. For an overview of graph-based models used to design algorithms for wireless networks we refer to Ref. [76].

Coloring a general graph is not only an NP-Complete problem, but is also hard to approximate to within factor of $n^{1-\epsilon}$, for any constant $\epsilon > 0$ [89]. Wireless networks, however, can usually be better modeled by more restricted classes of graphs, such as *geometric graphs*. Geometric graphs are graphs whose vertices are placed in a metric space (usually in a two-dimensional Euclidean plane), and two vertices are connected if and only if the distance between them is less than or equal to some radius r, for some $r > 0$. When $r = 1$, the geometric graph is commonly called a Unit Disk Graph (UDG) (see Figure 6.1). When the radius is different for each node and two vertices u and v are connected if and only if the distance between them is less than or equal to the minimum of the two radii, then the graph is called a *disk graph*. Intuitively, disk graphs are intersection graphs of (possibly equal sized) circles in the plane and have been extensively used to model broadcast networks.

In Ref. [14] it was proved that a series of closely related problems to wireless scheduling, such as coloring in graphs, independent set, domination, independent domination, and connected domination, are also NP-complete in UDGs. Interestingly enough, finding cliques when a geometric representation (circles in the plane) of a UDG is provided, was shown to be polynomial in time.

Although the general graph model is too pessimistic, as the connectivity of most networks is not arbitrary but obeys certain geometric constraints, the UDG model (or other disk models) is too optimistic. In reality, radios are not omnidirectional, and the presence of obstacles often impacts connectivity. In heterogeneous environments, such

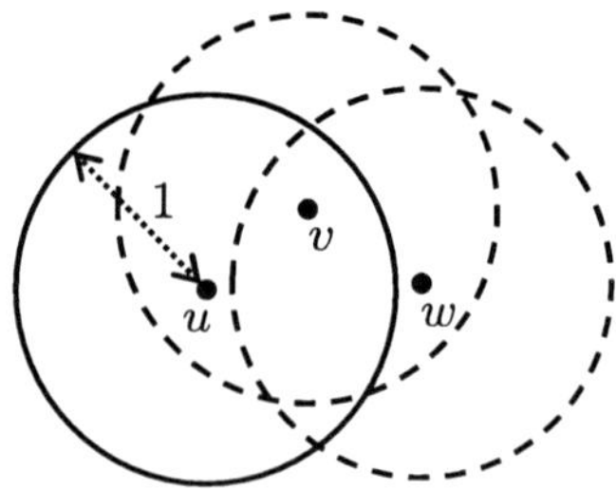

Fig. 6.1 Unit disk graph: node u is adjacent to node $v(d_{vu} \leq 1)$, but not to node $w(d_{vw} > 1)$.

as inner-city or in-building networks, a node might not be able to communicate to nodes which are close by, but located in a room across the wall. Nevertheless nodes within the same room are typically highly connected. This scenario suggests that in highly obstructed regions, the number of neighbors which are not adjacent is small. An interesting model to represent this kind of scenario is the so-called *bounded independence graph*, or BIG. In this model, if $\Upsilon^d(v)$ denotes the set of independent (or pairwise non-adjacent) nodes which are at most d hops away from node v in the connectivity graph G, G is said to have bounded independence iff $\forall v \in G, |\Upsilon^d(v)| = O(poly(d))$, where $poly(d)$ is a function polynomial in d, i.e., $poly(d) = d^{O(1)}$. Note that the UDG model is a special case of the BIG model.

One commonly used graph-based interference model is the protocol model [38]. In this model, a transmission by a node s_i is successfully received by a node r_i iff the intended receiver r_i is sufficiently apart from the sender s_j of any other simultaneous transmission, i.e., $d(s_j, r_i) \geq (1 + \rho)d(s_i, r_i), \forall s_j \neq s_i$. The constant $\rho > 0$ models situations, where a guarding region is specified by the protocol to prevent a neighboring node from transmitting (on the same channel) at the same time. This model implicitly assumes that senders use power control to adjust their signals. There are, therefore, two radii: a transmission range R_T and an interference range R_I. A node can successfully transmit to a receiver node in its transmission range only if the receiver is not within the interference range of any other concurrently transmitting node (see Figure 6.2).

Another group of graph-based interference models are the so-called k-hop interference models. In these models, no two links within k hops can successfully transmit at the same time. These models are even less realistic that the protocol model, since they overlook some crucial interference terms. For example, a $(k + 1)$-neighbor can be close to the receiver (see Figure 6.3).

The inefficiency of graph-based scheduling protocols in the physical interference model has been shown both through simulations [7, 35, 36], and through experiments (on mica2 sensor nodes running with TinyOS) [67]. In fact, in Ref. [67], Moscibroda et al. show that any protocol which obeys the laws of graph-based models can be broken by

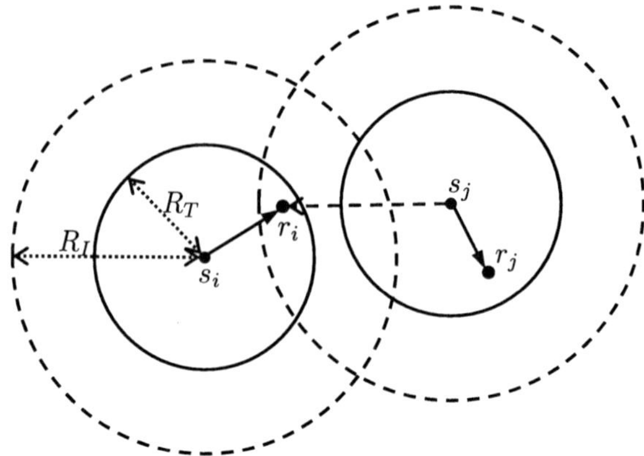

Fig. 6.2 Protocol interference model: there are two radii: transmission range R_T and interference range R_I. In this example, node r_i is not able to receive a transmission from node s_i if node s_j concurrently transmits to node r_j — even though r_i is not adjacent to s_j.

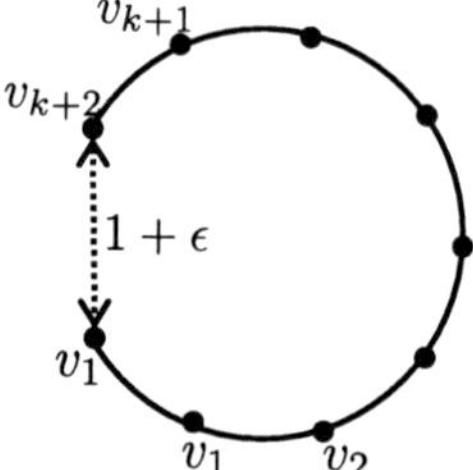

Fig. 6.3 Example where k-hop interference fails: nodes v_1 and v_{k+2} are separated by a path of $k + 1$ hops, but are close (distance $1 + \epsilon$).

a protocol explicitly defined for the physical model. There have been some efforts to model the properties of the physical model using SINR-derived conflict graphs, e.g., [11, 47, 78], however, the obtained bounds are usually too loose, or are only valid in restricted network topologies.

6.2　Other Physical Models

In this monograph, we have used the physical model. In wireless communication research, this is only one of many available fading channel models.

In many digital wireless communication standards, e.g., Wireless LAN 802.11 or UMTS, a higher signal-to-noise ratio can be used to employ an advanced digital modulation to allow for higher data rates. As such, the 802.11 g standard can for instance achieve up to 54 Mbit/s if conditions are perfect, and will slow down if base station and client hardly hear each other. As a rule of thumb, the achievable bit rate will be about logarithmic in the signal-to-(interference-plus)-noise ratio. This is a natural extension to the model and work presented in this monograph.

One may also extend the physical model by modeling obstructions more accurately. This leads to physical ray tracing models where the architecture of the situation is represented in the system. As in graphical ray tracing, physical ray tracing systems will compute shadowing, reflection, scattering, or diffraction of wireless signals by walls or other physical objects. Clearly, one could imagine studying for instance wireless scheduling in these physical ray tracing models.

Finally, there are various statistical wireless channel models, such as Rayleigh or Rician fading models. These fading models assume that the magnitude of a signal will vary randomly, according to a stochastic process. In Rayleigh fading, for instance, the magnitude of a signal varies as a radial component of the sum of two uncorrelated Gaussian random variables. In wireless scheduling, one may hope to opportunistically use long-term fading components, i.e., to transmit when conditions are good.

More generally, network information theory considers a set of senders and receivers, and a channel transition matrix describing the effects of interference and noise in the network [15]. Network information theory wants to determine how much information can be transmitted; in general, apart from special cases, this is an open problem.

For more information on wireless channels, we refer to text books on wireless communication fundamentals, e.g., [80]. For cross-layer aspects of modeling wireless networks, such as the interaction of the physical layer with the network and the transport layers, we refer to the survey by Georgiadis et al. [29].

7

Conclusions

In this monograph we presented several selected results on algorithmic wireless scheduling in the physical model — and beyond. This line of research is still young, at the time of writing there are more open than solved problems. In Section 3 we presented wireless scheduling without power control, a subject that is reasonably well understood. Nevertheless, there is a multitude of open problems, for instance whether polynomial time approximation schemes (PTAS) are possible. The problem of scheduling with power control, analyzed in Section 4, is even a bit less understood, although very recently there was a breakthrough result [53] featuring an algorithm with an approximation ratio independent of the network topology. If we go beyond scheduling (Section 5), one can easily list hundreds of important open problems. One can study scheduling in wireless multi-hop networks, or combine it with routing, to understand the capacity of arbitrary wireless networks. One can study specific traffic, such as broadcast or convergecast. Virtually any problem ever studied in the context of wireless networks gets a new flavor in this context. There is no doubt that some of earlier knowledge may be transferable, especially when dealing with higher-layer problems. But as the example of wireless scheduling shows, in lower layers of the network stack, techniques may be quite different. Finally, as sketched in Section 6, one may look into more general interference models. Hopefully, this line of research leads eventually to new, more efficient protocols.

References

[1] R. Ahlswede, N. Cai, S. Y. R. Li, and R. W. Yeung, "Network information flow," *IEEE Transactions on Information Theory*, vol. 46, no. 4, pp. 1204–1216, 2000.

[2] E. Ásgeirsson and P. Mitra, "On a game theoretic approach to capacity maximization in wireless networks," *Arxiv preprint arXiv:1008.1556*, 2010.

[3] C. Avin, Y. Emek, E. Kantor, Z. Lotker, D. Peleg, and L. Roditty, "SINR diagrams: Towards algorithmically usable SINR models of wireless networks," in *Proceedings of the ACM Symposium on Principles of Distributed Computing (PODC)*, pp. 200–209, 2009.

[4] C. Avin, Z. Lotker, F. Pasquale, and Y. A. Pignolet, "A note on uniform power connectivity in the SINR model," in *Proceedings of the International Workshop on Algorithmic Aspects of Wireless Sensor Networks (ALGOSENSORS)*, 2009.

[5] C. Avin, Z. Lotker, and Y. A. Pignolet, "On the power of uniform power: capacity of wireless networks with bounded resources," in *Proceedings of the Annual European Symposium on Algorithms (ESA)*, 2009.

[6] S. Banerjee and A. Misra, "Minimum energy paths for reliable communication in multi-hop wireless networks," in *Proceedings of the ACM International Symposium on Mobile Ad Hoc Networking and Computing (MOBIHOC)*, pp. 146–156, 2002.

[7] A. Behzad and I. Rubin, "On the performance of graph-based scheduling algorithms for packet radio networks," in *Proceedings of the IEEE Global Telecommunications Conference (GLOBECOM)*, 2003.

[8] V. Bhandari and N. H. Vaidya, "Capacity of multi-channel wireless networks with random (c, f) assignment," in *Proceedings of the ACM International Symposium on Mobile Ad Hoc Networking and Computing (MOBIHOC)*, pp. 229–238, ACM, 2007.

[9] V. Bhandari and N. H. Vaidya, "Connectivity and capacity of multi-channel wireless networks with channel switching constraints," in *Proceedings of the IEEE International Conference on Computer Communications (INFOCOM)*, pp. 785–793, May 2007.

[10] S. A. Borbash and A. Ephremides, "The feasibility of matchings in a wireless network," *IEEE/ACM Transactions on Networking (TON)*, vol. 14, no. SI, pp. 2749–2755, 2006.

[11] G. Brar, D. Blough, and P. Santi, "Computationally efficient scheduling with the physical interference model for throughput improvement in wireless mesh networks," in *Proceedings of the ACM International Conference on Mobile Computing and Networking (MOBICOM)*, 2006.

[12] D. Chafekar, V. Kumar, M. Marathe, S. Parthasarathy, and A. Srinivasan, "Cross-layer latency minimization for wireless networks using SINR constraints," in *Proceedings of the ACM International Symposium on Mobile Ad Hoc Networking and Computing (MOBIHOC)*, 2007.

[13] C.-N. Chuah, D. Tse, J. Kahn, and R. Valenzuela, "Capacity scaling in MIMO wireless systems under correlated fading," *IEEE Transactions on Information Theory*, vol. 48, no. 3, pp. 637–650, 2002.

[14] B. N. Clark, C. J. Colbourn, and D. S. Johnson, "Unit disk graphs," *Discrete Mathematics*, vol. 86, no. 1–3, pp. 165–177, 1990.

[15] T. Cover and J. Thomas, *Elements of Information Theory* vol. 1. Wiley Online Library, 1991.

[16] M. Dinitz, "Distributed algorithms for approximating wireless network capacity," in *Proceedings of the IEEE International Conference on Computer Communications (INFOCOM)*, 2010.

[17] M. Dinitz and M. Andrews, "Maximizing capacity in arbitrary wireless networks in the SINR model: Complexity and game theory," in *Proceedings of the IEEE International Conference on Computer Communications (INFOCOM)*, 2009.

[18] O. Dousse, F. Baccelli, and P. Thiran, "Impact of interferences on connectivity in ad hoc networks," *IEEE/ACM Transactions on Networking (TON)*, vol. 13, no. 2, p. 436, 2005.

[19] T. ElBatt and A. Ephremides, "Joint scheduling and power control for wireless ad-hoc networks," in *Proceedings of the IEEE International Conference on Computer Communications (INFOCOM)*, 2002.

[20] T. Erlebach and T. Grant, "Scheduling multicast transmissions under SINR constraints," *Algorithms for Sensor Systems*, pp. 47–61, 2010.

[21] A. Fanghänel, S. Geulen, M. Hoefer, and B. Vöcking, "Online capacity maximization in wireless networks," in *Proceedings of the ACM Symposium on Parallelism in Algorithms and Architectures (SPAA)*, pp. 92–99, 2010.

[22] A. Fanghänel, T. Kesselheim, H. Räcke, and B. Vöcking, "Oblivious interference scheduling," in *Proceedings of the ACM Symposium on Principles of Distributed Computing (PODC)*, 2009.

[23] A. Fanghänel, T. Kesselheim, and B. Vöcking, "Improved algorithms for latency minimization in wireless networks," in *Proceedings of the International Colloquium on Automata, Languages and Programming (ICALP)*, 2009.

[24] M. Fussen, R. Wattenhofer, and A. Zollinger, "Interference arises at the receiver," in *Proceedings of the International Conference on Wireless Networks, Communications, and Mobile Computing (WIRELESSCOM)*, 2005.

[25] F. Gantmacher, *The Theory of Matrices* vol. 2. Chelsea Publishing Company, 1959.

[26] Y. Gao, J. C. Hou, and H. Nguyen, "Topology control for maintaining network connectivity and maximizing network capacity under the physical model," in *Proceeding of the IEEE International Conference on Computer Communications (INFOCOM)*, 2008.

[27] M. R. Garey and D. S. Johnson, *Computers and Intractability, A Guide to the Theory of NP-Completeness*. W. H. Freeman and Company, 1979.

[28] M. Gastpar and M. Vetterli, "On the capacity of wireless networks: The relay case," in *Proceedings of the IEEE International Conference on Computer Communications (INFOCOM)* vol. 3, pp. 1577–1586, 2002.

[29] L. Georgiadis, M. J. Neely, and L. Tassiulas, "Resource allocation and cross-layer control in wireless networks," *Foundations and Trends in Networking*, vol. 1, no. 1, pp. 1–144, 2006.

[30] O. Goussevskaia, M. Halldórsson, R. Wattenhofer, and E. Welzl, "Capacity of arbitrary wireless networks," in *Proceedings of the IEEE International Conference on Computer Communications (INFOCOM)*, 2009.

[31] O. Goussevskaia, T. Moscibroda, and R. Wattenhofer, "Local broadcasting in the physical interference model," in *Proceedings of the ACM International Workshop on Foundations of Mobile Computing (DialM-POMC)*, Toronto, Canada.

[32] O. Goussevskaia, Y. A. Oswald, and R. Wattenhofer, "Complexity in geometric SINR," in *Proceedings of the ACM International Symposium on Mobile Ad Hoc Networking and Computing (MOBIHOC)*, pp. 100–109, September 2007.

[33] O. Goussevskaia and R. Wattenhofer, "Complexity of scheduling with analog network coding," in *Proceedings of the ACM International Workshop on Foundations of Wireless Ad Hoc and Sensor Networking and Computing (FOWANC)*, May 2008.

[34] S. A. Grandhi, R. Vijayan, D. J. Goodman, and J. Zander, "Centralized power control in cellular radio systems," *IEEE Transactions on Vehicular Technology*, vol. 42, no. 4, pp. 466–468, 1993.

[35] J. Grönkvist, "Interference-based scheduling in spatial reuse TDMA," PhD thesis, Royal Institute of Technology, Stockholm, Sweden, 2005.

[36] J. Grönkvist and A. Hansson, "Comparison between graph-based and interference-based STDMA scheduling," in *Proceedings of the ACM International Symposium on Mobile Ad Hoc Networking and Computing (MOBIHOC)*, pp. 255–258, 2001.

[37] P. Gupta and P. R. Kumar, "Critical power for asymptotic connectivity in wireless networks," *Stochastic Analysis, Control, Optimization and Applications: A Volume in Honor of W. H. Fleming*, pp. 547–566, 1998.

[38] P. Gupta and P. R. Kumar, "The capacity of wireless networks," *IEEE Transactions on Information Theory*, vol. 46, no. 2, pp. 388–404, 2000.

[39] M. Haenggi and R. K. Ganti, "Interference in large wireless networks," *Foundations and Trends in Networking*, vol. 3, no. 2, pp. 127–248, 2009.

[40] M. Halldórsson, "Wireless scheduling with power control," in *Proceedings of the Annual European Symposium on Algorithms (ESA)*, pp. 368–380, 2009.

[41] M. Halldórsson and P. Mitra, "Wireless capacity with oblivious power in general metrics," in *ACM-SIAM Symposium on Discrete Algorithms (SODA)*, 2011.

[42] M. Halldórsson and R. Wattenhofer, "Wireless communication is in APX," in *Proceedings of the International Colloquium on Automata, Languages and Programming (ICALP), Track C*, 2009.

[43] J. Hamkins, "Joint Viterbi algorithm to separate cochannel FM signals," in *Proceedings of the IEEE International Conference on Acoustics, Speech, Signal Processing*, 1998.

[44] J. Hamkins, "An analytic technique to separate cochannel FM signals," *IEEE Transactions on Communications*, vol. 48, no. 4, pp. 543–546, 2000.

[45] R. Hekmat and P. van Mieghem, "Interference in wireless multi-hop ad-hoc networks and its effect on network capacity," *Wireless Networks*, vol. 10, pp. 389–399, 2004.

[46] Q.-S. Hua and F. C. Lau, "Exact and approximate link scheduling algorithms under the physical interference model," in *Proceedings of the ACM International Workshop on Foundations of Mobile Computing (DialM-POMC)*, pp. 45–54, New York, NY, USA: ACM, 2008.

[47] K. Jain, J. Padhye, V. N. Padmanabhan, and L. Qiu, "Impact of interference on multi-hop wireless network performance," in *Proceedings of the ACM International Conference on Mobile Computing and Networking (MOBICOM)*, pp. 66–80, New York, NY, USA: ACM Press, 2003.

[48] R. M. Karp, "Reducibility among combinatorial problems," in *Complexity of Computer Computations*, pp. 85–103, 1972.

[49] S. Katti, S. Gollakota, and D. Katabi, "Embracing wireless interference: Analog network coding," in *Proceedings of the SIGCOMM*, pp. 397–408, ACM Press, 2007.

[50] S. Katti, I. Maric, A. Goldsmith, D. Katabi, and M. Medard, "Joint relaying and network coding in wireless networks," in *Proceedings of the IEEE International Symposium on Information Theory (ISIT)*, pp. 1101–1105, June 2007.

[51] B. Katz, M. Völker, and D. Wagner, "Energy efficient scheduling with power control for wireless networks," in *Proceedings of the International Symposium on Modeling and Optimization in Mobile, Ad Hoc, and Wireless Networks (WiOpt)*, pp. 160–169, 2010.

[52] T. Kesselheim, "Packet scheduling with interference," Master's thesis, RWTH Aachen, Germany, 2009.

[53] T. Kesselheim, "A constant-factor approximation for wireless capacity maximization with power control in the SINR model," in *ACM-SIAM Symposium on Discrete Algorithms (SODA)*, 2011.

[54] T. Kesselheim and B. Vöcking, "Distributed contention resolution in wireless networks," in *Proceedings of the International Symposium on Distributed Computing (DISC)*, pp. 163–178, 2010.

[55] D. R. Kowalski and M. A. Rokicki, "Connectivity problem in wireless networks," in *Proceedings of the International Symposium on Distributed Computing (DISC)*, pp. 344–358, 2010.

[56] U. C. Kozat and L. Tassiulas, "Throughput capacity of random ad hoc networks with infrastructure support," in *Proceedings of the ACM International Conference on Mobile Computing and Networking (MOBICOM)*, pp. 55–65, New York, NY, USA: ACM, 2003.

[57] G. Kramer and S. Shamai, "Capacity for classes of broadcast channels with receiver side information," in *Proceedings of the IEEE Information Theory Workshop (ITW)*, pp. 313–318, 2007.

[58] V. Kumar, M. Marathe, S. Parthasarathy, and A. Srinivasan, "Algorithmic aspects of capacity in wireless networks," in *Proceedings of the International Conference on Measurement & Modeling of Computer Systems (SIGMETRICS)*, 2005.

[59] E. Lebhar and Z. Lotker, "Unit disk graph and physical interference model: Putting pieces together," in *Proceedings of the IEEE International Parallel and Distributed Processing Symposium (IPDPS)*, 2009.

[60] T.-H. Lee, J.-C. Lin, and Y. T. Su, "Downlink power control algorithms for cellular radio systems," *IEEE Transactions on Vehicular Technology*, vol. 44, 1995.

[61] S. Li, Y. Liu, and X.-Y. Li, "Capacity of large scale wireless networks under Gaussian channel model," in *Proceedings of the ACM International Conference on Mobile Computing and Networking (MOBICOM)*, pp. 140–151, New York, NY, USA: ACM, 2008.

[62] Z. Lotker and D. Peleg, "Structure and algorithms in the SINR wireless model," *SIGACT News*, vol. 41, no. 2, pp. 74–84, 2010.

[63] P. Monks, V. Bharghavan, and W. W. Hwu, "A power controlled multiple access protocol for wireless packet networks," in *Proceedings of the IEEE International Conference on Computer Communications (INFOCOM)*, pp. 1567–1576, 2001.

[64] T. Moscibroda, "The worst-case capacity of wireless sensor networks," in *Proceedings of the International Conference on Information Processing in Sensor Networks (IPSN)*, pp. 1–10, 2007.

[65] T. Moscibroda, Y. A. Oswald, and R. Wattenhofer, "How optimal are wireless scheduling protocols?," in *Proceedings of the IEEE International Conference on Computer Communications (INFOCOM)*, 2007.

[66] T. Moscibroda and R. Wattenhofer, "The complexity of connectivity in wireless networks," in *Proceedings of the IEEE International Conference on Computer Communications (INFOCOM)*, April 2006.

[67] T. Moscibroda, R. Wattenhofer, and Y. Weber, "Protocol design beyond graph-based models," in *Proceedings of the ACM SIGCOMM Workshop on Hot Topics in Networks (HotNets)*, 2006.

[68] T. Moscibroda, R. Wattenhofer, and A. Zollinger, "Topology control meets SINR: The scheduling complexity of arbitrary topologies," in *Proceedings of the ACM International Symposium on Mobile Ad Hoc Networking and Computing (MOBIHOC)*, 2006.

[69] J. O'Rourke, "Advanced problem 6369," *American Mathematical Monthly*, vol. 88, no. 10, p. 769, 1981.

[70] A. Ozgur, O. Leveque, and D. Tse, "Hierarchical cooperation achieves optimal capacity scaling in ad hoc networks," *IEEE Transactions on Information Theory*, vol. 53, no. 10, pp. 3549–3572, 2007.

[71] S. Pillai, T. Suel, and S. Cha, "The perron-frobenius theorem: Some of its applications," *IEEE Signal Processing Magazine*, vol. 22, no. 2, pp. 62–75, 2005.

[72] B. Rankov and A. Wittneben, "Achievable rate regions for the two-way relay channel," in *Proceedings of the IEEE International Symposium on Information Theory (ISIT)*, pp. 1668–1672, July 2006.

[73] B. Rankov and A. Wittneben, "Spectral efficient protocols for half-duplex fading relay channels," *IEEE Journal on Selected Areas in Communications*, vol. 25, no. 2, pp. 379–389, February 2007.

[74] T. Rappaport, *Wireless Communications: Principles and Practice*. Upper Saddle River, NJ, USA: Prentice Hall PTR, 2001.

[75] A. Richa, P. Santi, and C. Scheideler, "An O(log n) dominating set protocol for wireless ad-hoc networks under the physical interference model," in *Proceedings of the ACM International Symposium on Mobile Ad Hoc Networking and Computing (MOBIHOC)*, 2008.

[76] S. Schmid and R. Wattenhofer, "Algorithmic models for sensor networks," in *Proceedings of the International Workshop on Parallel and Distributed Real-Time Systems (WPDRTS)*, April 2006.

[77] S. Singh and C. S. Raghavendra, "PAMAS–power aware multi-access protocol with signalling for ad hoc networks," *SIGCOMM Computer Communications Review*, vol. 28, no. 3, pp. 5–26, 1998.

[78] K. Sundaresan, W. Wang, and S. Eidenbenz, "Algorithmic aspects of communication in ad-hoc networks with smart antennas," in *Proceedings of the ACM International Symposium on Mobile Ad Hoc Networking and Computing (MOBIHOC)*, pp. 298–309, New York, NY, USA: ACM, 2006.

[79] TIK Computer Engineering Group, *Sensor Network Museum*. 2010. (accessed September 13, 2010), http://www.snm.ethz.ch.

[80] D. Tse and P. Viswanath, *Fundamentals of Wireless Communication*. Cambridge University Press, 2005.

[81] M. Vu, N. Devroye, M. Sharif, and V. Tarokh, "Scaling laws of cognitive networks," in *Proceedings of the International Conference on Cognitive Radio Oriented Wireless Networks and Communications*, pp. 2–8, 2007.

[82] K. Wang, C. Chiasserini, R. Rao, and J. Proakis, "A joint solution to scheduling and power control for multicasting in wireless ad hoc networks," *EURASIP Journal on Applied Signal Processing*, 2005.

[83] J. E. Wieselthier, G. D. Nguyen, and A. Ephremides, "Energy-efficient broadcast and multicast trees in wireless networks," *Mobile Networks and Applications (MONET)*, vol. 7, no. 6, pp. 481–492, 2002.

[84] F. Xue and P. Kumar, "The number of neighbors needed for connectivity of wireless networks," *Wireless Networks*, vol. 10, no. 2, pp. 169–181, 2004.

[85] F. Xue and P. R. Kumar, "Scaling laws for ad hoc wireless networks: An information theoretic approach," *Foundations and Trends in Networking*, vol. 1, no. 2, pp. 145–270, 2006.

[86] J. Zander, "Distributed cochannel interference control in cellular radio systems," *IEEE Transactions on Vehicular Technology*, vol. 41, no. 3, pp. 305–311, 1992.

[87] J. Zander, "Performance of optimum transmitter power control in cellular radio systems," *IEEE Transactions on Vehicular Technology*, vol. 41, 1992.

[88] S. Zhang, S. C. Liew, and P. P. Lam, "Hot topic: Physical-layer network coding," in *Proceedings of the ACM International Conference on Mobile Computing and Networking (MOBICOM)*, pp. 358–365, 2006.

[89] D. Zuckerman, "Linear degree extractors and the inapproximability of max clique and chromatic number," in *Proceedings of the ACM Symposium on Theory of Computing (STOC)*, 2006.